P9-DET-351

EVOLUTION
ISN'T WHAT IT USED TO BE

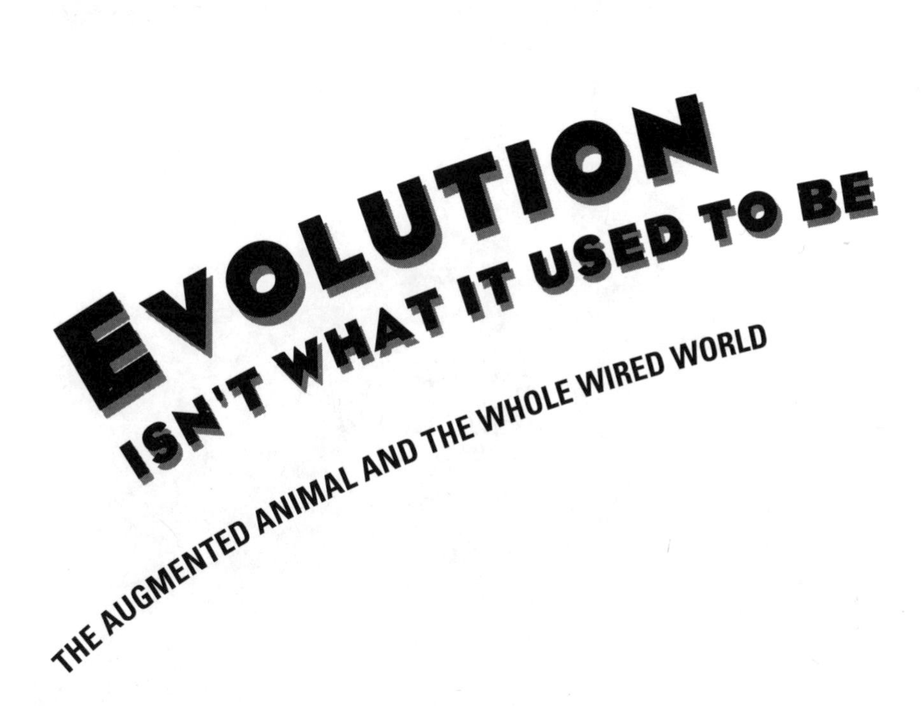

Evolution Isn't What It Used to Be

The Augmented Animal and the Whole Wired World

Walter Truett Anderson

W. H. Freeman and Company
New York

For Dan
and his world

Library of Congress Cataloging-in-Publication Data

Anderson, Walt, 1933–

Evolution isn't what it used to be: the augmented animal and the whole wired world / Walter Truett Anderson.

p. cm.

Includes bibliographical references and index.

ISBN 0-7167-2998-9 (hard cover)

1. Human evolution. 2. Evolution (Biology) 3. Technological innovation. 4. Biotechnology. I. Title.

GN281.4A53 1996

573.2—dc20

Printed in the United States of America

First printing 1996,VB

Jacques Ellul, in his concern with the negative aspects of technology, suggests that "its evolution . . . is progressing almost without decisive intervention by man." There is a larger and perhaps unintentional truth in this observation. Technology may be as natural a part of human evolution as the differentiation of finger and thumb, and in this sense has been until now almost as free from man's possible control. We do not yet fully understand or accept the organic evolutionary quality of technological growth. The idea is, in itself, somewhat alien to our comprehension. . . .

As created, renewed, and ultimately directed by human life, technology is as organic as a snail shell, the carapace of a turtle, a spiderweb, or the airborne dandelion seed. In many respects, it is now more ubiquitous as a functional component of the ecosystem than any organic life form other than man himself. The amounts of energy converted by machines, the materials extracted from the earth, processed, recombined, and redistributed in the technological metabolism, and the gross effects of such increased metabolic rates on the ecosystem, are now greater than the effects of many global populations of other organic species. Guidance and control of our technological and other recent systems can be realized only through greater understanding of their growth. So far, there has been little effort to understand; more attention has been given to unheeding exploitation or equally unreflective rejection.

—John McHale

Contents

INTRODUCTION:

What if Everything Changed and Nobody Noticed?

I think it is a good idea to offer your basic argument at the beginning of a work such as this. So I will state mine here, as clearly as I can: it is that we are in the midst of a profound evolutionary transition—happening right now, within us and all around us. We are becoming a different kind of animal from any that has existed before. We are also living in a different kind of world—and inhabiting it in a different way, unlike other species and unlike humans of a few generations past.

As this transition unfolds, we, and the world we live in, and all life on it, are changing in fundamental and irreversible ways. People are busily unraveling the human genetic code, and learning how to alter it, and altering the genetic makeup of other species, and creating new kinds of living things. This is happening now, and—despite the sound and fury about biotechnology in some circles—it is happening with surprisingly little excitement. The convergence of technologies is a big part of this, but the really profound event of our time is not that technologies are converging: it's that we are converging with our technologies. Our minds are augmented by computers and other devices, and our bodies are augmented by a wide range of evolutionary inventions, from vaccines to artificial organs—and they are truly parts of what we are. Meanwhile the world is becoming a bionic planet—remodeled, wired, and networked by information systems that monitor its health, forecast its future, and govern its ecosystems. And the world is not separable from those information systems. All these things are happening now, and are certainly not secrets, yet hardly anybody seems to have noticed. Everybody has noticed some of the parts, of course, but few people have put them together.

What you get when you put them together is a true evolutionary transition, one of those strange leaps that life makes when it becomes something entirely new and different—when everything changes. My job in this book is to convince you that this is happening—and, furthermore, that it is happening much more quickly than the major evolutionary transitions that have taken place in the past, and that we will not be able to continue not noticing.

My job, also, is to convince you that this evolutionary transition is a practical, down-to-earth sort of event, one that presents concrete problems and opportunities to all kinds of people. Right now, governments are making policy decisions about the world's climate and atmosphere. Businesses are succeeding or failing on the basis of their skill at managing fast-breaking changes in medicine and agriculture. Individuals and families are dealing with new choices, new powers, new ways to solve some of their problems and new things to worry about. The transition presents real and pressing ethical issues and—I think even more urgent—enormous equity issues, because there is reason to fear that the new information resources and the amazing devices that augment human life are being distributed even less fairly than more ordinary goods like food and shelter.

I have mentioned technological convergences, which are important forces in the transition; in a sense, it is made of convergences. The word *convergence* is in the air these days, and most of the excitement is about the coupling of electronic technologies such as computers and television. That is indeed hot stuff, but now people are beginning to notice another one, sometimes called the "bionic" convergence—the convergence of the bio-sciences with all the electronic technologies. This is the subject we'll explore in the first chapter, because it leads us toward comprehending what John McHale called the "organic evolutionary quality of technological growth"—its inseparability from our own minds and bodies, from the evolution of our species.

The bionic convergence has been under way for some time, because the new biotechnologies (like our kids) grew up in the age of computers. There would be no recombinant DNA, no Human Genome Project, no genetic therapy opening a new chapter in medical history, without computer technology. Global environmental monitoring—our fundamental tool for understanding and coping with global concerns such as the greenhouse effect—is also a child of the computer era. Today the bio-sciences and information/communications technologies (ICT) are racing ahead hand in hand, and as they do they jointly transform personal life and global politics in exciting, promising, worrisome ways.

Individuals are discovering that they have a bewildering range of new options about managing their bodies, and the human species finds itself managing the biosphere.

I will offer some predictions about where all this is taking us. Among them:

1. ***Getting used to eugenics:*** The word *eugenics*—meaning conscious efforts to improve the quality of human genetic inheritance—has acquired a taboo status since the time of Hitler, and for good reason. Nonetheless, I will show that a new era of eugenics is beginning, and that this *de facto* eugenics is both more and less than what people usually have in mind when they use that fearful word.

2. ***Rethinking ecology:*** Environmentalism today is devoutly and somewhat mindlessly antitechnological, with a deep romantic yearning for the past. As a lifelong environmentalist I have strong emotional links to this kind of thinking, but it is severely, dangerously limited. Environmentalists are generally very good at noticing problems—and their movement has been valuable for that—but their deep-seated fear of the future prevents them from fully grasping what is happening in the world. I believe the future will belong to a new kind of proactive environmentalism that harnesses ICT and embraces biotechnology.

3. ***Greening industry:*** In the early 1900s a group of social philosophers—including Lewis Mumford—speculated about a shift to a "biotechnic" society, in which biological production systems would replace the inorganic machines of the factory. Something of that sort is indeed on the horizon. Many industrial processes (from mining to manufacturing) will be altered by biotechnology, and new biomaterials will be created.

4. ***Moving into the bio-information society:*** As you put the various pieces together—the far-reaching information/communications hookups, the progress of the Human Genome Project, the international system of gene banks containing the seeds of millions of plant and animal species, the satellites monitoring ecosystems from space—you can't help glimpsing a global social order in which a vast bio-information network surrounds the Earth, and affects virtually everything we do. This new society will require different structures of governance, different life skills and values. This new social order will be the product of human creation, but we can't simply build it the way you can draw up plans and construct a building; we have to learn our way into it.

Nobody can prove whether a prediction will come true, of course. But I can offer strong evidence to show that these predictions aren't nearly as far out as they may appear at first glance. In fact, it's cheating a little to call them predictions, because each of the things I'm talking about is already happening.

It is remarkable that there is so little public discussion of the present evolutionary transition. There is much loose talk about great forward leaps in New Age circles—too much, in my opinion—and there are quite a few people who see the significance of the bionic convergence and are trying very hard to find (or invent) the language to describe their visions to the rest of us. But there is far too little talk about such matters in the humdrum worlds of governance and business, or at the individual and family level—in the places where the transition is guided, shaped, and lived. Insofar as the pieces of the transition are dealt with explicitly, it is exactly that—as pieces. We do not seem to have a public vocabulary for putting the pieces together—telling ourselves the larger story as a way of understanding where we fit in it as individuals and how much of it we can write as a society.

I believe there is much to be gained by stretching for the larger story, using it to frame our thinking about public and private issues. And there is much to be gained by actively, thoughtfully, creatively participating in this transition—especially since nobody is going to get to *not* participate in it.

PART ONE

THE BIONIC CONVERGENCE

CHAPTER ONE

The Computer Meets the Gene

bi-on´-ic, *adj. utilizing electronic and mechanical devices to assist humans in the performance of difficult, dangerous, or intricate tasks; bio- + (electro)nic.*

But now, something new is on the scene, which is we actually have the ability to do experiments on evolution. We can, within the computer, run populations of hundreds of thousands of generations or even millions of generations, and watch the process of evolution. . . .

There's more: There is this guy who is evolving proteins that do RNA catalysis. He generates a whole bunch of random RNA and then arranges for them to bind if they are capable of doing this catalysis. Then he filters them so the ones that do the right thing are overrepresented in the mixture. And then he amplifies these few with DNA techniques and he harvests a generation, and then repeats that. At the end of it he gets very specific proteins that do specific things. So he gets evolution in a test tube, literally, now. There's no reason why this process couldn't be automated somehow. That's cool—it's like being around when they were making the first transistors.

—Danny Hillis[1]

Evolution is far, far more than that distant Darwinian business, the thing scientists argue about and fundamentalist theologians argue against. It is the growth and change of all living things, of life itself, and it is central to everything that happens to our bodies and to our minds and to the Earth on which we live. It is the subject that orchestrates all other subjects. It is—as Teilhard de Chardin, the theologian who got himself in trouble for seeing it as the hand of God, wrote—"a light illuminating all facts, a curve that all lines must follow."[2]

Evolution has many mysterious, fascinating aspects; the most mysterious and fascinating of all is that evolution evolves. That life changes is amazing in itself—but the way life changes also changes. Occasionally the process invents a new way of moving ahead, lifts itself to another level, commences a new stage of growth. We can see that such leaps have occurred in the past. We know, for example, that a marvelous, unpredictable, irreversible transition occurred when replicating molecules first combined to become something like living organisms. They changed, and the world changed. We know that such a transition happened again, millions of years later, when the descendants of those first organisms invented the systems of symbolic communication we call speech.[3] The talking apes became a different sort of species, and the world, by virtue of having such creatures in it—creatures that could build fires, domesticate animals, plant seeds in the ground, survive in all climates—became a different sort of world. We can see these things looking back, we can even imagine other such transitions in the future—the science-fiction writers are great at that—but we have a hard time believing such a transition is happening right now.

Yet we are surely in the midst of one. It is not as easy to describe, because there are so many facets to it, but it has to do with an explosion of human capacity to study and manipulate life, and to affect the course of evolution. One of the most important parts of this transitional process is the convergence of the biological revolution with the information revolution, of biology with electronics.

You don't hear much yet about the convergence of the biotechnologies with the information and communication technologies, but it is likely to capture the public's attention soon. There was a time when nuclear physics monopolized the public's imagination concerning the far reaches of knowledge—$E = mc^2$, the Manhattan Project, and all that—and then there was a time a few decades later when the exploration of space took over and tended to shape the common perception of what science and/or technology could do. Neither of those has entirely lost its grip—we are still dazzled by photons and quarks, still use "rocket scientist" as a synonym for a really smart dude. But for sheer mind-boggling intellectual enterprise—coupled with an astonishing speed of movement from abstract concepts to practical application—nothing quite compares to what is happening on the frontiers of biology, at the meeting place of two amazing information systems: the computer and the gene.

From the Butterfly Net to the Internet

The dawning of the bio-information era is, like so many major historical events, an emergent. It is something greater than the sum of its parts, and therefore unpredictable—but now, looking back, we can see that it has been on the way for a long time. It was unpredictable, and it was also inevitable.

The bionic convergence has been in gestation for centuries, because the study of life has always been the product of convergences. People who are in search of new understanding of such things as plants and animals, or microscopic life, or the ecology of large systems always use whatever mechanical and intellectual tools happen to be available at any given time. Biologists are always either hindered or helped by the state of technology, including information technology, and by the state of progress in other fields of knowledge such as mathematics.

Aristotle, one of the pioneer biologists (the first person we know of who tried to do a systematic classification of all the different kinds of living things), had a limited number of species available to study. He couldn't study what he couldn't see, and his work was limited by the absence of microscopy; millions and millions of life forms were literally invisible. It was limited in another way by the boundaries of a known world that was more or less the equivalent of the Mediterranean basin. What lay beyond those boundaries was also not visible, and most information about it was produced by the wild imaginary flights of people who had never been there. The methods for recording and disseminating information were also primitive by our standards. Aristotle and his students had to record their findings on fragile papyrus scrolls. The first book—the handwritten and hand-bound parchment pages the Romans called the codex—had not yet appeared. Probably not more than a few hundred people read Aristotle's studies in his time, and much of his work has been lost forever.

When the Swedish biologist Carl von Linnaeus set to work in the eighteenth century, things were much different. Thanks to progress in mapmaking, shipbuilding and navigational technology, the world had become round; travelers could cross the oceans and observe the flora and fauna of distant lands. Many of Linnaeus's students did just that, and a fair number of them died from exotic diseases in the process. Thanks to progress in typesetting and printing, it was possible for Linnaeaus's work to be published in books and distributed to scholars. A wider circle of data-gathering; more people in the loop; faster transmission.

Charles Darwin did his own field work in his famous voyage aboard the H.M.S. *Beagle,* and could have communicated his findings to other students of natural science much more quickly than he did if he had not played around with his data for over twenty years afterward. But when *On the Origin of Species by Natural Selection* was finally published in 1859, it sent a shock wave through the world—a slow-moving one by current standards, but nonetheless impressive. By the time of Darwin's death—another twenty years later—the *Origin* was being translated, published, tested, written about and preached against everywhere. No scientific discovery before had been able to make such an impact in such a short time.

The mathematical approach to evolutionary science—which soon became a central and essential part of the field—did not begin with Darwin. He was never much of a mathematician. But his cousin Francis Galton was a nut about statistics, the original data nerd. Galton believed *anything* could be measured. He even applied mathematics to religious questions that he thought testable. Did prayer increase human longevity? Were ships carrying good God-fearing missionaries less likely to go down at sea? While he was exploring these heady subjects, Galton also began compiling statistics on inherited characteristics in human beings. His work laid the foundation for a new branch of natural science, biometrics, that began to emerge after the discovery of the almost-lost experiments of Gregor Mendel.

Mendel, prelate of the monastery at Brunn, was in some ways a rather worldly and learned man. He had studied mathematics, physics, and biology at the University of Vienna and later he explored a number of other subjects including meteorology. He was a careful scientist, and he analyzed tens of thousands—perhaps hundreds of thousands—of samples in his breeding research. And he knew of Darwin's work: a copy of the *Origin* in German, marked and obviously much read, was found in Mendel's library after his death in 1884. But he was still only an amateur, not really a part of the scientific world, and he had no direct communications with other scientists. He read a paper on his work to the local natural history society, but the members apparently were not greatly stimulated by Father Mendel's adventures in pea breeding. They did, however, duly publish the paper. (They had to; he was the founder of the society.) Undoubtedly Charles Darwin would have been fascinated by the 1866 *Transactions of the Brunn Natural History Society*—containing the paper that many biologists consider as significant as the *Origin*—but he never found out about it. Probably he would have if the report had been published in one of the established scientific journals of the time, but the bio-information system was still quite rudimentary, and Darwin did not get the news from Brunn.

Darwin was not the only person who failed to hear about Mendel's work; just about everybody failed to hear about it for many years. The Brunn Society had sent out 120 copies of its 1866 *Transactions.* Mendel's paper was cited in a few other obscure publications, and then it disappeared from view until, in 1900, three different researchers—one in Germany, one in Austria, one in the Netherlands—discovered it independently of one another, recognized its importance, and wrote the papers that made Mendel—posthumously—a famous man.

One of the three codiscoverers was a Dutch botanist, Hugo de Vries. *His* paper was published in a scientific journal and read—quite soon after its publication—by the zoologist William Bateson in England. Bateson found in it exciting evidence of the *particulate* nature of inheritance—the idea that a plant's or animal's characteristics are not simply a blend of the characteristics of its parents, but instead result from the workings of some other mechanism (nobody knew precisely what it might be) that operated in a more complex fashion.

By this time networks of communication—including several new journals—had been established among the far-flung community of scientists and theorists who were trying to build on Darwin's work. Among the other forces that were influencing the course of events in this post-Darwinian scientific world were advances in technology—notably the microscope—and attempts to apply greater mathematical precision to natural science.

The new microscopes—with stronger lenses, glass slides, and artificial dyes—had enabled Walther Flemming in Germany to study within animal cells the small particles that came to be called chromosomes, from the Greek word for color, because they absorbed the dye used in microscopy. (New language was being invented along with new instruments.) Some scientists, building on Flemming's work, were beginning to suspect that the chromosomes might contain the messengers of heredity, the very keys to evolution.

The synthesis of mathematical thinking with Darwinian theory and laboratory research might have proceeded much more quickly than it did if it had not been handicapped by petty conflict and academic backbiting among the leading natural scientists of the time. But almost in spite of themselves, they moved toward a general recognition that there had to be some "units of inheritance"—people began calling them *genes,* from the Greek word for "to produce"—within all living things.

Through the first half of the twentieth century the young science of genetics—a word Bateson had invented to go with genes—grew slowly but steadily. Much of the post-Darwinian work on evolution was done in laboratories, as researchers compiled mountains of data on the inheritance of specific characteristics. One of the most industrious of these researchers was Thomas Hunt Morgan of Columbia University, whose favorite object of study was the fruit fly *Drosophila melanogaster*. Its charm lay in its ability to thrive on simple food in simple surroundings—Morgan generally used milk bottles—and in its short life cycle. In ten days' time a fruit fly would hatch, mature, and start producing still more fruit flies. With such short generations, you could gather a lot of data in ten years of research, and, from about 1907 to 1917, Morgan did. He was not idly collecting numbers, but systematically tracking mutations and refining Mendelian theory—and searching for the gene. The term was being widely used, although it was still a very abstract concept. A gene was one of those things, like black holes in space, that had to be there. But it was not seen or understood; only its effects were seen.

A familiar sight in Morgan's fly room, besides the bottles, was a device he and his students had invented to help them keep track of their work. This gadget, of which they were enormously proud, was simply a four-sided wooden box—each side representing one of *drosophila's* chromosomes. On each side of it were thumbtacks, which they would move horizontally and vertically to indicate progress in locating a trait. That was Morgan's computer.

Morgan's work, carried on by others, led eventually to the first chromosome maps, which identified the location of specific genes. But the gene remained elusive. In some ways it became even more elusive as the work progressed, as scientists all over the world shared their experimental findings. The more they found out, the more clearly they could see that the gene's workings were damnably complex. It was some submicroscopic entity that carried instructions for all the characteristics of a living organism, and it could behave in curious and ever-surprising ways. Even when they had a pretty good idea of where a certain gene was located, they were still mystified as to what it was.

The prevailing assumption was that it had to be a protein. Protein molecules were known to be large, complex, and capable of doing amazing things such as carrying oxygen in the blood and enabling chemical combinations to take place. But Oswald Avery, a shy and quiet American scientist who studied bacteria in his laboratory at the Rockefeller Institute in New York (bacteria were becoming as popular as fruit flies, since they could produce a new generation in less than a half hour), gradually came to the conclusion that the "transforming principle," at least in the strains of pneumonia bacteria he was studying, was not protein but deoxyribonucleic acid. "Who could have guessed it?" he wrote to his brother. Max Delbruck, the biophysicist, said later: "At that time, it was believed that DNA was a *stupid* substance. . . ."[4]

It took some years from the time of Avery's discovery in 1944 for the scientific community to accept the astonishing idea that the DNA molecule was not a stupid substance, but an elegant information system capable of carrying all the instructions for the gestation of a living organism—and that, furthermore, it was the information system for nearly all life on Earth except for a few viruses based on the closely related RNA. By the time the doubters came around, the race was already on to figure out how in the world DNA did what it did. Its chemical components—the four bases plus phosphate and deoxyribose—were known by 1952, and so were the atomic connections that brought them together to form the molecule. But the geneticists knew there was something they didn't know—the actual three-dimensional structure of the molecule—and they suspected that this had to be the key to DNA's remarkable information-bearing ability.

A technology that had been developed by physicists, X-ray diffraction analysis, eventually made it possible to figure out the structure. Using this method, a researcher could produce a sort of photograph—a bewildering Rorschachlike splatter of dots that told little or nothing to the naked eye but could be analyzed mathematically—to get an idea of how a biological molecule was arranged in space. It was an enormously exacting and time-consuming method, practiced by only the most patient of scientists. James Watson and Francis Crick, the two men who finally discovered DNA's famous double-helix structure in 1953, had both worked on X-ray analysis of proteins before they set to work trying to build a model of the molecule. An X-ray study of DNA that had been done by a Cambridge colleague, Rosalind Franklin, provided the final piece of information that completed the model, made them famous, and launched a revolution.

The actual model was an angular structure that they made themselves (using a system Linus Pauling had developed in America) by soldering together bright pieces of tin and copper wires. It represented the way the DNA molecule is shaped—a breathtaking discovery in itself—and also showed how it works, how the four bases function as an alphabet to write the gene's instructions. Watson and Crick, and other scientists who came to view their metal sculpture, thought it was not only instructive but beautiful and deeply moving, an amazingly graceful way for nature to carry out its most important task.

Word of the discovery spread quickly from Cambridge. The first formal publication of the Watson-Crick finding was in *Nature,* but many scientists had already heard the news before the article came out. Some were skeptical, of course. By the time the new view of DNA became generally accepted in the biological world, it had leaped beyond the boundaries of science into the mass media. Watson and Crick were interviewed everywhere, written about in the newspapers and magazines, talked about on the TV shows. Watson even turned up—photographed with Richard Burton—in the pages of *Vogue.*

Once the basic concept of the genetic text had been accepted, the next step was to begin reading some parts of it. Another Cambridge scientist, Fred Sanger—who had won a Nobel Prize for his work with protein molecules—took the lead in this. Sanger was a hands-on kind of investigator, who had an enormous capacity for painstaking laboratory work and an equally great distaste for publicity. He had to develop entirely new research techniques for working with DNA, and it took years of effort before complete sequences of DNA could be read. But as the technique developed and was picked up by other researchers, the whole enterprise gained speed. By 1970 dozens of DNA characters in various biological systems had been read; then in 1977 the number jumped quickly into the thousands; in the 1980s it went into the millions, then quickly into the tens of millions. The 1977 leap was marked by the publication—again in *Nature*—of the first complete genome, the entire text of a virus called phi-X-174.

By the time that first complete genome had been sequenced, Stanley Cohen and Herbert Boyer in California had already figured out how to snip a piece of genetic information from one organism, transfer it to another, and get the host organism to "express" the new gene—that is, carry out its instructions. By 1982 the first commercial product based on this recombinant DNA technology—human insulin manufactured by bacterial fermentation—was on the market.

As gene studies progressed, biology began to outgrow its favorite medium of communication, the printed scientific journal. A single gene contains anywhere from 2,000 to a million nucleotide sequences; the best medium—indeed, the only really workable medium—for these huge compilations of data was the computer. When Watson and Crick first created their DNA model, computers had been of utterly no interest to most biologists. Crick once tossed off a contemptuous reference to them, saying, "It is better to use one's head for a few minutes than a computing machine for a few days."[5] But as genome analysis proceeded, so did the biological-electronic convergence.

Both of the two major repositories of human genome data—GenBank at the Los Alamos National Laboratory in New Mexico and the European Molecular Biology Laboratory in Heidelberg, Germany—were state-of-the-art centers when launched, but they had to be enlarged some 25 times in the first four years of operation. Despite that, *Science* reported in 1986 that they were swamped.

At about this time, the whole field of molecular biology was becoming computerized—not only computerized, but *personal*-computerized. At first, only a few large companies could afford the expensive hardware and special programming that were needed for computerized analysis of biological data. But in the 1980s, with prices coming down and new products becoming available, virtually any laboratory could utilize the computer's speed and power. Software companies were enthusiastically designing new programs for biologists, and entrepreneurs were launching start-up companies that aspired to make a buck selling biological computer and data services. Existing data banks were converting their information, new ones were being established. A new science, bio-informatics, was becoming recognized as a vital intellectual link between the information/communications revolution and the biological revolution, and new facilities were springing up everywhere. In 1989 the U.S. National Library of Medicine established a Center for Biotechnology Information, aimed at speeding up the flow of knowledge between researchers, and from researchers to people creating medical applications. One of its star features was a smart access system called GenInfo that enabled researchers to ask questions in their own words and have them answered by information retrieved instantly from a dozen different databases. In 1995 the European Bioinformatics Institute was opened near Cambridge as an outstation of the ever-expanding information needs of genetics researchers.

The launching of the Human Genome Project, with its ambitious goal of producing a record of the entire sequence of genes in human DNA, had of course given a tremendous boost to this bio-informatizing boom. So had the creation of the Internet.

The Internet is easily the fastest-growing communications system ever created, and some enthusiasts have no trouble at all calling it the greatest of all human inventions. Not the least amazing thing about it is that it was originally started by the U.S. government—first as a kind of virtual think tank for the Pentagon, then, after being taken over by the National Science Foundation, as an information link for research and education. Nobody ever really planned for it to develop into such a huge global network. As a puckish *New Republic* writer put it, the whole phenomenon shows how much the government can accomplish when it doesn't put its mind to it.

Words can describe the Internet and the World Wide Web in a general way, but you really can't write about the whole system in great detail or accuracy for two reasons. One is that it contains more information than any library. The other is that it grows more quickly than anyone can write, print, and publish a book or article about it.

At the moment, the Internet serves as the world's electronic highway, connecting the ever-growing horde of web sites and data bases. Actually, "highway" is a rather feeble metaphor for a communications system with so many linkages, such capacity to make geography seem irrelevant. No doubt it will evolve, be replaced by still bigger and faster systems with greater compatibility among its various parts and better data transfer capacity. But its present global reach and speed are impressive enough. It does the job for tens of millions of scholars, students, bureaucrats, hackers, and geeks—and it is creating a global scientific information network quite unlike anything that has previously existed. So the advent of computers and information technology doesn't just facilitate the work of the individual researcher; it creates an entirely new working environment for science as a whole—and particularly for genetics, the science of biological information.

Genetic research has been totally transformed by a development that absolutely nobody foresaw at the time of the initial Watson-Crick breakthrough. Modern genetics simply could not have developed as it has apart from its technological base, any more than modern literature could have evolved without the printing press. The information system has indeed become a kind of super-brain whose memory contains the accumulated findings of a worldwide network of scientists. And in the process, the computer has also become the metaphor of choice for describing the gene. No longer is DNA a kind of book. Instead, science writer M. Mitchell Waldrop—author of *Complexity*—calls it "a kind of molecular-scale computer," that directs the cell's work as it builds itself and repairs itself and interacts with the outside world.[6]

Biology Goes into Cyberspace

At first, the value of computer technology was that it offered a way to store and communicate data. That's an important function—it can hardly be overestimated, since the computer is the only man-made information system that comes anywhere near the complexity of the DNA molecule—but it is still only a part of the story.

As software becomes more elegant and hardware less expensive—and as the campuses and the laboratories are taken over by a new generation of biologists who grew up playing video games on home computers—information technology becomes a part of the research process itself, seamlessly integrated into the work and thought of science. Computers perform measurements and computations, test hypotheses, prepare research materials. Some of them use color and graphics to help scientists think about genetic activity—much as the wooden box and the thumbtacks helped T. H. Morgan, and the beautiful heap of metal helped Watson and Crick.

A whole range of automated research assistants is now available, beguilingly advertised in full color in the science magazines. Want to create some copies of a gene sequence by means of polymerase chain reaction? Buy a RoboCycler System from Stratagene, and you don't even have to pay extra for rights to use the patented technique: "When you buy a RoboCycler System, Stratagene will give you the PCR authorization at no charge!" Want to do some gene sequencing? Buy the new sequencer from Perkin Elmer: "You've never seen a system this fast. Using simultaneous multicolor detection and advanced electrophoresis technology, the ABI PRISM 377 generates data four times faster than the ABA 373." Want to hunt for mutations? Buy the D GENE system from Bio-Rad. "No experience is required to be an efficient mutation hunter! Until now these techniques have been difficult to perform due to cumbersome equipment. The D GENE system is an instrument system breakthrough which allows these powerful methods to be performed simply, even by novices."[7]

A touch closer to everyday reality for most of us is another sort of biology-information synthesis, the use of computers to study diseases and surgical techniques, visualize the workings of the human body. The First World Congress on Computational Medicine, Public Health and Biotechnology was held in Austin, Texas, in April of 1994. Its preliminary announcement proclaimed: "Computational instruments are now used, not only as exploratory tools but also as diagnostic and prognostic tools. The appearance of high performance computing environments has, to a great extent, removed the problem of increasing the biological reality of the mathematical models. For the first time in the history of the field, practical biological reality is finally within the grasp of the biomedical modeler."[8]

Biomedical imaging and modeling are making rapid strides right now—as rapid as the eighteen-month doubling time of the speed of computers. Institutions such as the National Laboratory of Medicine are developing new data-visualization programs that are more or less the physiological equivalent of the Human Genome Project. You hear about the Visible Human Project, the Visible Embryo Project, the Digital Anatomist Program. They are making it possible to see inside the human body with greater detail and precision than anybody has ever done before, and in the process they are creating new ways of practicing and teaching medicine.

The Visible Human Project (also known as the Adam and Eve project) took thousands of photographs of minutely dissected cross-sections of two cadavers—one male, one female—and then converted this material into digitized data. It can be brought up on the computer screen and viewed from different angles and at different levels of detail, as if the viewer were a Superman with X-ray vision.

Such models allow students and teachers to study any organ, take it apart and (something you can't do with a cadaver) put it together again. They can put a tumor into the body and watch it grow, observe the inner effects of diseases and accidents. Similar imaging and modeling techniques enable surgeons to visualize a specific portion of a patient's body—such as a damaged pelvis or skull—and design computer-generated bone replacement sections. The next generation of Visible Humans will probably be completely lifelike supercomputer-generated simulations in which the blood flows, the muscles contract, and the bones and organs move.

Then there's evolutionary modeling. Computers are great devices for creating abstract visual models, electronic representations of the real world—lifelike beings that can be viewed from different angles, that act and move, change over time, respond to events in their environment. As the theory and technology of modeling advance, we begin to see marvelous pieces of computer art/science such as the electronic wonder called *Tierra,* which Thomas Ray at the University of Delaware created to explore Darwinian theory.

Tierra is a world—a computer world resembling the solid world of millions of years back. It is populated by digital organisms programmed to live and replicate—and to strive to make more copies of themselves. In this world great complexity emerges. The programmed creatures grow and mutate, make mistakes that lead to evolutionary change. They replicate, diversify. Some lose the ability to replicate, yet survive as viruslike parasites by borrowing instructions from others that have been more successful in climbing up the evolutionary ladder.

Scientists seek to gain from *Tierra* some new understanding of the most fundamental mysteries of life: how complex plants and animals might have evolved from simple submicroscopic organisms, organisms that had evolved out of the interaction of chemicals in the primordial chaos of 4 billion years or so in the past.

The technique of combinatorial chemistry (also known as directed molecular evolution, although it might more accurately be called selective breeding of molecules) is another dramatized enactment of the real thing, only staged in a more traditional sort of laboratory and using molecules of RNA or DNA instead of computer programs. The molecules behave in some ways like complete organisms, and they don't take up much room—even less than bacterial cultures. This makes it possible for the investigators to work with huge populations—an experiment may involve 10 trillion molecules swimming about in a concentrated solution. Submitted to different pressures and stimuli, they evolve—they increase their ability to perform certain basic chemical tricks, and even learn entirely new ones. Their main disadvantage is their relatively slow rate of producing new generations—two days or so—but the scientists who are working with this piece of high-tech Darwinism think they may be able to speed it up to about fifty generations a day with new technology. This research of course requires computers to keep track of the trillions of performers. And the researchers are a team of colleagues who work together—together intellectually, but not geographically. They do their work in laboratories separated by thousands of miles, but they are linked by computers and electronic data transmission.

Like many of the fields of research evolving out of the biotech-infotech convergence, combinatorial chemistry is both pure science and applied science. The pure science is after a more precise understanding of how evolution actually takes place. Researchers are looking calmly at the prospect of achieving one of the favorite feats of the science-fiction writers, creating life in a test tube. One of them told a science reporter: "If we can get to the point where, instead of us directing the evolution by adding in protein enzymes to help the replication occur, the RNAs were selected to do their own replication, then it's no longer directed evolution, then it's self-sustained evolution. Some people might call that life."[9]

The applied science pursues new pharmaceutical products, biochemicals that are able to target diseases such as multiple sclerosis. It is contributing to a medical revolution that has unfolded so quickly that most doctors still don't quite know what has happened.

Shouting at the Doctors

Late in 1993 *The New York Times* reported that the medical world was being subjected to an extraordinary barrage of scientific articles—more than 150 different reports on advances in molecular genetics in eleven medical journals—orchestrated by the officers of its own professional association. The news story said: "Convinced that advances in molecular biology and molecular genetics are transforming the theory and practice of medicine, the American Medical Association is virtually shouting at the nation's doctors to pay attention." One of the ringleaders of the project called the outpouring of articles "an explosion, a culmination, a profound statement of the importance of this subject for the human condition."[10]

The AMA was trying to tell its members that everything important to doctors and patients—knowledge about the causes of disease, methods of treatment, legal issues, ethical issues, everything—is being transformed by genetic discoveries. One of the articles claimed that not since the Middle Ages, when a few researchers began to edge toward the discovery of bacteria, "has the stage been set for as radical a rethinking of the causes of disease and disability as is now occurring in conjunction with the elucidation of the human genome." Another pointed out that although the changes already visible were astonishing, they were only the smallest fraction of what was likely to unfold: "We are," reported the author, "really only beginning to see the tip of the iceberg."

The Iceberg Cometh

I'll close this brief review of biological/informational happenings with a prediction: in the decade just ahead—the last years of the twentieth century, the early years of the twenty-first—the world will see a cascading sequence of revolutions in biological science and technology.

The first to claim the public's full attention will probably be the one that the AMA is currently getting exercised about—the revolution in medicine. Actually there will be several of those, and their cumulative effect will be a major transformation of diagnosis and treatment and health maintenance—with corresponding changes in the rules of personal life for all of us.

Unfolding at the same time will be a revolution in agriculture, an event no less dramatic and momentous than the invention of farming ten thousand years ago. This will bring forth new foods, new ways of producing food, new variations on old foods, new nonfood agricultural products (such as medicines), enormous upheavals in international trade, an entirely new set of approaches to economic development, and arguments galore.

Meanwhile, a bio-industrial revolution will alter many kinds of manufacturing and produce new sources of energy and chemical feedstocks—and at the same time will create entirely new industries and render some present industries obsolete. Along the way fortunes will be built, and undoubtedly some fortunes will be lost by people and companies that failed to pay attention to the news from the laboratories. Most important, these revolutions will not happen once and then be done: they will be *ongoing* revolutions, building on themselves.

Although all this may sound a bit breathless, it is not a particularly risky prediction. If anything, it is on the conservative side. It is scarcely a prediction at all, because each of the revolutions I have mentioned above is already well under way. Any competent scientist in the various fields we call biotechnology will tell you that breakthroughs are being made at an astonishing rate. Most competent scientists, however, are also specialists. They know what is happening in their own fields, but few put the pieces together. This book is about putting the pieces together.

Some of the pieces have to do with what we call biotechnology. Biotechnology *is* information. This ought to be quite obvious—but it hasn't been quite fully grasped yet, either by those who seek to establish exclusive ownership over some piece of biotechnology or by those who dream of stamping out the whole enterprise.

Some of the pieces also have to do with ecology, and with the study of the biosphere—because information technology is having an enormous impact on how we think about life on Earth. It transforms not only what scientists do, but what all of us talk about and worry about. Ecological information has taken a central place in politics, and is a major force in shaping a new global society.

This new global society is a *bio-information society,* created by a coming together of such pieces—a convergence of sciences and technologies including biotechnology, earth sciences, ecology, and information/communications technologies.

In the following chapters we will look at the various pieces of this convergence, commuting between the globe and the gene, the macro and the micro, with stops at a few midpoints between the two. Let's start at the micro level—with biotechnology, the troublesome surge of activity that flowered so quickly out of genetic research and is now beginning to reveal its world-changing potential.

CHAPTER TWO

Microbionics: A Gathering of Revolutionaries

We are now, though we only dimly begin to realize the fact, in the opening stages of the Biological Revolution—a twentieth-century revolution which will affect human life far more profoundly than the great Mechanical Revolution of the nineteenth century or the Technological Revolution through which we are now passing.

—Gordon Rattray Taylor[1]

I go to a lot of conferences about biotechnology issues, and it is a regular feature of such events—you can depend on it—that sooner or later in the proceedings somebody from a biotech company takes the podium and begins by saying that biotechnology is not really new at all. Why, hell, people have been doing it for thousands of years. Way, way back there in ancient times they were using bacteria to make yogurt and cheese, ferment beer.

This is literally true, but somewhat disingenuous. It is correct that fermentation falls within the standard definition of biotechnology. It is true that people were using bacteria effectively before the dawn of history—even though people didn't really know until the middle of the nineteenth century that bacteria existed.

But the comparison suggests that nothing has changed, and that is a serious manifestation of the not-noticing syndrome. Something very important *has* changed. Fermentation when you know you are using bacteria is different from fermentation when you put a bunch of stuff in a pot and something funny happens to it. Fermentation when you know the genetic code of the bacteria is different still. And fermentation when you have for all practical purposes created the bacteria —and are using it to cook up a batch of human growth hormone—is very different indeed.

People are now using bacteria (and other micro-organisms, and other plants and animals) in entirely new ways, and these new uses sprout forth from an ever-growing base of genetic knowledge. This explosion of information is in itself one of the most rapid and spectacular evolutionary transitions ever to have taken place on the planet, and we would be kidding ourselves if we regarded it as anything less than that. In practice it is not one biological revolution, but several of them happening at once.

What It Isn't, What It Is

The biological revolution—by which I mean the various forms of biotechnology based on modern genetics and on the bio-info convergence—has not happened before. It is *like* some things that have happened before—the industrial revolution, the invention of agriculture, and, yes, the discovery of fermentation—but it is truly *sui generis,* and opens up a new chapter in human history.

This hasn't really become clear yet. People are having a hard time figuring out what it is that has been figured out. Much of the public discussion of biotechnology consists of attempts to define it by comparing it to something else. The dialogue is partly in the language of science, but it is also in metaphors that compete to say what the new science most *resembles.* The metaphors are generally chosen for their usefulness in advancing the political agendas of the speaker.

Those who have taken up arms against biotechnology like to compare it to dangerous organic chemicals—the herbicides and pesticides that Rachel Carson wrote about—and to nuclear power, which promised us clean energy and gave us Chernobyl. The agenda they advocate is to impose strict regulations upon all uses of biotechnology or, better yet, regulate it out of existence. The favorite metaphor is putting the genie back in the bottle.

People from the industry prefer to compare biotechnology to the other industries—such as automobiles and electronics—in which America enjoyed an early competitive advantage but later fell behind. Their favored agenda, of course, is to create a friendly climate in which the industry will flourish and maintain its present lead over foreign competitors. And it's very easy, particularly in the business-happy United States, to equate the science with the industry. Even the biotech-bashers are sometimes inclined to do that, because it implies the whole purpose of biotechnology is to make money. That really doesn't do, however, as a way of getting a handle on what's happening. It's quite true that there are people trying to get rich off of biotechnology. There are also people trying to get rich off of religion—sometimes with better luck than those who are trying to get rich off of biotechnology—but in neither case can the larger phenomenon simply be reduced to its economic exploitation.

The various comparisons and metaphors that have been invoked in the public debates about biotechnology are useful, but limited. If we are to understand the biological revolution and have any kind of a productive discussion about its importance and meaning in our lives, we are going to have to find a better definition—a more spacious and suitable framework. If we don't, we are likely to waste much time and energy on what management people call the "wrong problem problem."

Biotechnology is information of a specific and very interesting sort—human symbolic information (spoken, written, printed, entered into computers) about genetic information. This is information that was made possible by an evolutionary transition, the convergence of biology with new information/communications technologies. It simply did not exist in the world until fairly recently. As this information develops it produces new understanding—new insights into such matters as how evolution works, how characteristics are inherited, what goes on when a virus invades the body—and also new applications. People learn how to do things they couldn't do before, and do old things in new ways.

The New Medicines

One thing that people—people in the medical professions—can do is diagnose some diseases more quickly and with greater precision. Although diagnosis is probably the least-publicized use of biotechnology, it is the area in which it has had its most immediate impact. One reason for this is that diagnostic kits move through the U.S. federal approval process much more speedily than drugs or therapetic techniques.

One of the lesser-known biotechnologies—far less of an attention-getter than recombinant DNA but far more productive in the early years of the bio-information era—is the use of monoclonal antibodies. Antibodies are proteins the body manufactures as defenders. They mobilize against a foreign molecule (the antigen), lock onto it, and literally squeeze it to death. The remarkable thing about an antibody is its specificity. It precisely fits its enemy. An antibody against one kind of flu virus has no effect whatever against another kind. As one text puts it, "It is as if we had within our bodies an immense sleeping army of about 18 billion soldiers, each battalion of which only awakens at the approach of an enemy wearing a particular color of uniform."[2]

The basic structure of antibodies was determined in 1959. They obviously had many potential uses in research and therapy, but they were devilishly expensive and hard to come by in any useful quantity. Then in 1975 a team of researchers in England (Cesar Milstein and Georges Kohler) discovered a way to mass-produce them by fusing an antibody cell with a cancer cell—thus creating a hybrid cell (called a hybridoma) with the properties of both: it would produce the antibody and reproduce itself indefinitely. Milstein and Kohler thoughtfully neglected to patent their discovery, and a new science, technology, and industry sprang up practically overnight. Tens of thousands of new hybridoma lines were soon being developed each year, and in 1984 the Hybridoma Data Bank was established as a central information source on them.

Because of the specificity of antibodies, they lend themselves beautifully to diagnosis. And because diagnostic kits are used for *in vitro* testing (in medical laboratories) rather than *in vivo* pharmaceutical applications (inside human bodies), such a product doesn't have to go through so rigorous a course of safety testing. The manufacturer mainly has to show that it works. By the mid-1980s kits testing for human venereal diseases, hepatitis B, and other viral infections were in use, and observers were talking about a "revolution" in diagnosis.[3] This is the first of several times I will use the "r"-word in relation to the impact of different biotechnologies in different fields. More recent developments in this particular field have resulted in diagnostic kits for home use. That's another revolution, which we'll discuss in a later chapter because it has as much to do with *self-care* as with the standard forms of medical practice.

The bio-information convergence also makes it possible to do something no diagnostician could ever do in the past—tell you what disease you or your unborn child is going to have. The predictive capacity of genetic screening is limited—at the moment—to a fairly narrow range of problems, but it will inevitably expand as research progresses. This is one of the more ethically sensitive areas of biotechnology, because it means that diagnosis is far ahead of therapy: medical science can identify, often with great precision, genes for diseases nobody knows how to cure.

Recombinant DNA technology, perhaps the best known of the biotechnologies, enabled people to manufacture proteins for medical use by splicing human genes into bacteria. In its first stage this method was used to manufacture proteins (such as human insulin or growth hormone) that were more or less identical to those produced in the human body. The second stage, which is already here, is "protein engineering," in which the geneticists modify the proteins to make them more effective—or even create proteins that never existed before.

When I visited Japan to survey its progress in biotechnology in 1987, a huge new Protein Engineering Research Institute was just being launched, a multibillion-yen consortium financed by contributions from industries and the national government. Its purpose was to do basic research on the stunningly complex structure of proteins—basic research that would build up the data banks for future generations of protein architects.

Proteins, when you get right down to it, are pretty much the foundation of everything that living organisms are and do. They build our hair, our skin, our bones—and they also build the foods and fibers we harvest from domesticated plants and animals. They are the hormones that regulate our sex drives, the enzymes that digest our food, the antibodies that fight disease.

A protein is manufactured in the cells, under the direction of genes, and those directions can be modified. The technology for changing the instructions—protein engineering—is a highly promising source of treatments for many diseases, and it is a child of the bio-information convergence. One newspaper report aptly described it as "a cross between gene-splicing and computer modeling."[4]

We need to pause here and note that we are talking about a fundamentally new and different way of producing therapeutic drugs. Pharmacology has evolved through several stages in the past. First, people simply made use of things found in nature, such as medicinal plants. Later, they began to cultivate and harvest some of these plants. Later still—in fact quite recently—scientists learned how to synthesize the active ingredients. Currently, many new drugs are being discovered as technicians laboriously screen and test substances by the thousands and even millions. This is an increasingly sophisticated bio-information technology in itself: a company in Silicon Valley has developed a process that lets researchers synthesize different compounds on tiny chips of silicon and microscopic beads and then screen them for their medical potential.[5] Another company in the neighborhood, Shaman Pharmaceuticals, sends its researchers out around the world into distant villages and jungles to study the practices of native healers, searching for medically useful tropical plants whose compounds are then screened and studied with the latest technologies back in the home laboratory: another combination of old and new, another kind of global bio-information network.

Protein engineering is going through its own sequence of developments. First-generation protein engineering produces minor variations in existing proteins. We get a modified version of subtilisin (a bacterial protein used in detergents) that holds up in the presence of bleach; we get more effective forms of drugs such as beta interferon; we get industrial enzymes that work better at high temperatures. Second-generation "designer proteins" may be used in cancer therapy, water treatment, to make new plastics, textiles, and industrial substances. The third generation, which—depending on what you read—may never come or may be just around the corner, is nanotechnology: minuscule protein computers, submicroscopic protein machines that will sail through the bloodstream to fight disease or repair damage to the body. At the moment, we are somewhere in the first generation—perhaps inching into the second. But protein engineering is definitely here, and it is a bona fide medical revolution.

Meanwhile yet another medical revolution—gene therapy—is getting under way. This might seem to be the most obvious way to make use of genetic information. Scientists have known for some time that many serious illnesses are the result of a single defective gene. The list of these—over 200—includes muscular dystrophy, sickle cell anemia, cystic fibrosis, hemophilia, and a number of AIDS-like hereditary weaknesses of the immune system. One of the worst of those is severe combined immune deficiency, also known as the "bubble boy disease" because it results in an immune system so maimed that the victim must live in a protected environment—such as the plastic bubble in which one boy survived for twelve years until he finally died of an infection brought on by an attempted bone-marrow transplant.

Now, it seems logical enough that if genes are the cause of a problem, other genes might be the solution—that the perfect treatment would be to get some healthy DNA into the body so that whatever part of the system wasn't working would begin to function properly. Nevertheless, for many years the great majority of scientists dismissed the idea of gene therapy as impossible. Undoubtedly gene therapy would have overcome those doubts sooner or later and become a part of mainstream medicine. But it happened sooner, and that it was already being tried in the early 1990s is largely attributable to one very stubborn man, Dr. French Anderson.

French Anderson, now on the medical faculty of the University of Southern California, is a person who—as a magazine article about him put it some years ago—seems to have inherited a remarkable collection of genes himself.[6] Growing up in Tulsa, Oklahoma, he could read and write and do arithmetic before he started kindergarten, and was studying college textbooks by the time he was eight. Although a stutterer, he fought through that handicap with the same relentless determination he later showed in his professional life, and became a star member of his high school debating team. He also did well in athletics, won a scholarship to Harvard, and conceived the original idea of treating disease genetically while still an undergraduate. In a discussion with a visiting lecturer who was talking about the hemoglobin molecule (which transports oxygen in the bloodstream) and the sickle form of it that causes a severe anemia, he offered the suggestion that it might be possible to cure the illness by modifying the gene for hemoglobin. The lecturer contemptuously dismissed that idea, but a kind word from another professor encouraged Anderson to hang in there and see if such a thing could actually be done.

After completing Harvard Medical School, Anderson became a researcher at the National Institutes of Health in Washington, D.C., and it was there that he slowly developed the first gene-therapy experiments. His proposals for human trials were turned down with exhausting regularity by the NIH's Recombinant DNA Advisory Committee for several years. Then, in 1989, he finally won approval for the first test. This was not an attempt to cure a disease, but a nontherapeutic trial run with a terminally ill volunteer to find out whether a retrovirus (a virus made of RNA rather than DNA) could be used as a vector to take marker genes inside the human body and get them to express—that is, to direct a cell to manufacture a protein. Even that test was opposed by some members of the committee, and was further delayed when Jeremy Rifkin's Foundation on Economic Trends filed a lawsuit. The test was successful, however, and on September 14, 1990, Anderson and his colleagues proceeded to the first real trial of genetic therapy.

The patient was a five-year-old girl suffering from an inherited inability to make an enzyme called adenosine deaminase, or ADA. In the absence of ADA the body builds up a chemical that kills T-cells, which are central to the immune system. The girl was defenseless against any disease. A common cold was a major illness, and a case of chicken pox could be lethal. Since her parents discovered the nature of the disease, she had spent most of her life in seclusion. She was getting the best available treatment—weekly injections of a form of ADA derived from cows—but this did not completely restore her immune system.[7]

Had you been present at the historic event when the first genetic "operation" was performed, you would have seen what looked like a simple blood transfusion. It was anything but simple, however; it was the culmination of a highly complex procedure that had been developed over years of laboratory work and testing of its various components. Ten days earlier the scientists had taken several units of the girl's blood, and then removed the white cells and returned the red cells and plasma to her body. In a nearby laboratory, they had injected the white cells with a modified form of a mouse leukemia virus. The virus was to serve as the vector, the Trojan horse to get new genes into the girl's body. That's what viruses know how to do and what makes them dangerous: they invade the chromosomes of the host's cells like computer hackers and put the cells to work making more viruses. This virus had been rendered incapable of carrying out its own agenda, and instead carried the gene for human ADA. After ten days growing in a laboratory culture, the genetically treated white blood cells were returned to the girl's body in another transfusion—the first gene therapy "operation."

The next year Anderson and his colleagues treated another girl suffering from ADA deficiency. In May of 1993 both girls appeared at a press conference, happy and healthy. These treatments made it quite clear that gene therapy, once dismissed as an impossibility, is very possible indeed.

Gene therapy offers a whole lot more than treating victims of the two hundred or so single-gene diseases—although that is an amazing enough prospect in itself. It promises to become a commonplace method of therapy, not only for inherited illnesses such as cystic fibrosis, but also for other diseases such as some types of cancer. Late in 1994 a team in Massachusetts performed the first genetic "bypass operation"—injecting genes into the leg muscles of a man who was facing the possibility of an amputation because of severe blocking of an artery. The gene being injected was the one that codes for the production of a substance, called vascular endothelial growth factor (VegF), that can encourage new blood vessels to sprout out of an artery. Some tumors use VegF to grow new blood vessels to supply themselves with nourishment, and fetuses use it to build blood vessels; if this kind of treatment proves successful, it will be yet another revolution—this time in the treatment of vascular disease.[8]

Gene therapy will undoubtedly go through its own sequence of generations and refinements. A likely progression is from *in vitro* therapy, in which cells are removed from the body, genetically treated and then replaced, to tissue-specific *in vivo* therapy, in which healthy genes are introduced directly into the patient's cells; and from treatments that need to be repeated to treatments that provide a lifetime cure. The biggest transition, already exciting plenty of political and scientific controversy, will be the leap from somatic therapy—where the effects are not inherited—to germline therapy, which will modify genes in the reproductive cells.

A method called antisense technology—there's a name that biotech-bashers must love—opens up yet another approach to genetic therapy.

It is a method—in some ways the opposite of gene-splicing—for turning a gene off so it doesn't do what it normally would do. So far this approach is best known in connection with food. By blocking the gene in the plant that produces the ripening agent, it can prolong the shelf life of tomatoes—and probably other fruits and vegetables as well. But the same trick can conceivably be played on a sick cell that is producing a destructive protein in the human body. All diseases are basically the activities of malevolent protein molecules, and the body fights back by producing antibodies that bind to them and halt their action. Medicine generally fights back by using drugs that do roughly the same thing, but drugs sometimes attack other proteins as well and cause undesirable side-effects. Various researchers are busily trying to design antisense molecules that might become precise weapons against diseases such as viral herpes malaria, or might be used as a therapy to stop the proliferation of artery-hardening cells on the inner walls of blood vessels. And as they do, they watch for the latest news on the Human Genome front, because its data is the basic resource that makes sense of antisense.

Meanwhile, immunologists say we are about to see a lot of new vaccines: not only vaccines against diseases for which no immunization previously existed, but also new ways of delivering vaccines into the body—such as pills and nasal sprays, even bioengineered fruit.

There is not a single item on the above list of medical developments that has not been called revolutionary—and with good reason. Each of these revolutions will evolve and some of them will converge in various—and unpredictable—ways with other technologies such as laser surgery. And when the ground is shifting underneath so many areas of medical practice, from diagnosis to surgery, one would have to be conservative to the point of foolishness to say that nothing much is going on.

Behavioral Genetics

Then there's behavioral genetics with its view that human behaviors are determined by genetic inheritance. A few decades ago it was in disrepair and disrepute, as people turned against the racism, sexism, and plain bad science of its early enthusiasts. Now every day's news brings reports of research revealing genetic factors in manic-depressive disorder, some kinds of schizophrenia, alcoholism, even homosexuality. This is explosive stuff, because the arguments about the causes of various kinds of mental illness and personality characteristics are ancient and complex and bitter. They involve the vital interests not only of researchers and therapists but also of universities, funding agencies, governments, and the parents of mental patients—and, of course, of mental patients themselves.

The main battle lines have been drawn between two groups. On one side are those who favor a "medical model" of mental illness focusing on physical causes (such as evidences of chemical disturbance in the brain) and who look for physical remedies such as psychoactive chemicals. On the other side are those who emphasize the "psychosocial" causes, such as events in childhood, and who believe deeply in programs of therapy that focus on restoring patients' self-esteem and in some cases helping them develop new social and occupational skills so they can lead relatively normal lives. In recent years I have seen my colleagues in the humanistic psychology movement beat a steady retreat—not from their values, but from their total rejection of genetic analyses.

This doesn't mean behavioral genetics has been universally or uncritically embraced. As a writer in *BioScience* dryly observes, "The notion that human behavior is traceable to genes has not entirely lost its power to enrage."[9] What we have now is an ever-changing and enormously complicated argument about which genes play how much of a part in which behaviors. This is not, of course, merely an issue for the scientists. It has to do with how children are raised, how crimes are punished, how public funds are spent, how parents plan their families. And of course it affects the practice of psychotherapy. Already there is talk of genetic treatment—first for identified single-gene diseases such as Huntington's, perhaps eventually for more complex ailments such as manic-depression and Alzheimer's.

The New Agriculture

A virus-resistant potato is currently being field-tested in Mexico. The potato virus is an ancient destroyer of one of Mexico's major food crops—vitally important both to commercial growers and subsistence farmers. The addition of one gene in the potato creates an effective resistance to the virus. The commercial farmers do not then have to use chemical pesticides against the insects that carry the virus. The subsistence farmers—who can't afford chemicals—produce more food.

How do you get a potato to become virus-resistant? I asked Dr. Luis Herrera Estrella, the project director, about this, and he explained it to me as something like a microscopic form of birth control. What it comes down to is getting the gene for the virus coat protein—the wrapping around the viral DNA—into the potato. The virus has to shed its coat when it moves in on the host's chromosome. "So," said Dr. Herrera, "the virus comes to the potato. He wants to make love to the potato. He takes off his coat. Then—surprise!—the potato puts the coat back on."

Agricultural research is proceeding at a furious pace in laboratories all over the world—in universities, in chemical and seed companies, in publicly and privately funded research institutes. The U.S. Department of Agriculture has its own genome project, a ten-year study of the plant genes that code for such characteristics as drought resistance and effective use of nutrients. A worldwide network of plant biologists have been busy mapping and sequencing the complete genome of a flowering mustard weed, *Arabidopsis thaliana*—a plant that has no commercial value but is regarded by scientists as a good prospect for basic research because it has a relatively simple set of DNA. Along the way, some members of this group discovered the gene that makes one variety of the plant grow small cauliflowerlike curds instead of regular flowers. This discovery is expected to lead to improved varieties of commercial cauliflower (a relative of *Arabidopsis*) and probably to a whole range of new vegetables: bean cauliflower, corn cauliflower, almost-anything cauliflower.[10]

While the research data base on various crop plants builds, applications of all kinds are already being developed: improved nutritional qualities, disease and pest resistance, heat and drought tolerance, longer shelf life. The whole time scale of plant breeding is accelerating. At a conference on agricultural biotech, I recently heard a confident prediction that every major crop plant in the world will have been genetically modified in some degree by the end of this century. This means that in the next few years, more significant new varieties will be developed than have been developed in all the years from 1900 to the present time.

The New Bio-industries

The first industrial applications of biotechnology were in food processing (rennin produced by recombinant DNA is now widely used in making cheese) and pharmaceuticals. But other industries—ones we don't think of as "biological"—use enzymes as catalysts and detergents, and many more are likely to begin doing so. The European Biotechnology Information Service (in a newsletter article entitled "Toward the Greening of Industry") says: "Unlike fossil fuels, enzymes are renewable, potentially inexhaustible and easily carry out complex chemical reactions at room temperature and at atmospheric pressure. Furthermore, they can be improved by protein engineering and through genetic engineering and are available comparatively cheaply."[11]

Several new bio-industries are beginning to stir to life: bioremediation of environmental pollution, bio-mining, bio-materials, bio-energy, bio-electronics. It will not be long before these new industries and industrial processes begin to have significant impacts on human productivity, on economic thinking, and on the prosperity of many parts of the world.

The Case of the Disappearing Boundary

Everybody has heard about some of the things—good or bad—that people may do with biotechnology. The optimists tirelessly forecast scenarios of biological wonders in our future, while the technophobes, with equal stamina, paint pictures of monstrous disasters and periodically come down from the mountaintop with new instructions from God about what He did and did not have in mind for human beings to be able to do. Meanwhile the investors and speculators keep their eyes on the business pages for the latest news on whether biotechnology is a winner or a loser in the places where it counts—i.e., the stock markets.

All this has to do with the products of the biological revolution. And although it is perfectly understandable that people focus their attention on the tangible outputs—either those already on the market or those considered most likely to be developed—this is an inadequate way to understand any scientific revolution or any technology. Even when a line of research has reached the point of producing something that can legitimately be called an "invention," its inventors frequently haven't the foggiest notion of how it will ultimately be applied. The people at Bell Labs hesitated to apply for a patent on the laser because they couldn't see how it had any relevance to telephones. As it turned out, the laser did have all kinds of nontelephone uses—in navigation, precision measurement, chemical research, surgery (where it may ultimately render the scalpel obsolete)—and now, of course, it is finally revolutionizing the telephone industry as fiber-optic cables take over the connections around the world. Marconi, inventor of the radio, thought it would be useful as a kind of substitute telephone for "narrowcasting" in such uses as ship-to-shore communication; he had no idea of its potential for "broadcasting." It took people like David Sarnoff, an uneducated entrepreneur who came along later, to develop an industry that would do its part to transform the world.[12] Another classic in this category is IBM, which—even when it was successfully manufacturing and selling computers—did not guess at the vast market that would ultimately develop for personal computers such as the one which (in its present hybrid form as computer and typewriter) I'm now using to write this book.

The list goes on: knowledge finds its way around the world in unpredictable ways, new technologies combine with old ones (or other new ones), innovations occur that surprise even imaginative inventors. This process will inevitably repeat itself time and time again as we move farther into the bio-information age. What most people don't know is that there really isn't a clear boundary any more between biotechnology and plain biology. Biotechnological methods and knowledge are now commonly applied, for example, in identifying genetic traits of plants that are then developed through "conventional" breeding. We are dealing with a spread of new knowledge, new understanding, and new ways of doing things across all the life sciences.

So, although we commonly talk of biotechnology as "it"—as I have been doing up to this point and will probably lapse into doing again—there is really no such object that a small pronoun can accurately identify. "It" is a proliferation of new information about genetic information and how it works. "It" is not one new biotechnology but an ever-growing list of them: recombinant DNA, monoclonal antibodies, DNA probes, polymerase chain reactions, protein engineering, antisense technology, cell and tissue cultures. These techniques and variations on them are being used in laboratories all over the world. The more the information spreads, the less clear is the boundary.

What we have, then—and will have for some time to come—is an ongoing revision of the conditions and possibilities of life. The rules of the game will be rewritten, torn up, and rewritten again and again.

As new possibilities emerge, individuals are faced with new personal choices. And societies—including the global society that is now beginning to discover itself—are faced with new problems of governance and equity. Ethical dialogue comes alive again. Bioethics is yet another growth industry, with many people busily scanning our heritage of social and religious values in search of guidelines into a new and unfamiliar territory. No set of doctrines is entirely adequate to this changing information environment. We have to learn our way into it.

As we do, economic thinking changes as well. In the bio-information society, genetic information and the new kind of information we call biotechnology become the world's primary resources. Information reorders the whole hierarchy of economic and political values, because information is a fundamentally different kind of resource.

In later chapters I will expand on this, and show how the reordering is already beginning to happen. But first let's look at another kind of information, and at a different picture of what is currently going on in the world: the big picture.

CHAPTER THREE

Macrobionics: The Whole Wired World

Every fact that can be digitized, is. Every measurement of collective human activity that can be ported over a network, is. Every trace of an individual's life that can be transmuted into a number and sent over a wire, is. This wired planet becomes a torrent of bits circulating in a clear shell of glass fibers, data bases, and input devices.

—Kevin Kelly[1]

Migration, rather than stability in one place, is now the norm.

—Lilian Trager[2]

The astronauts' portrait of the planet Earth is one of the most reproduced photographs of all time. Shining forth from countless book covers, posters, and T-shirts, it has become a living piece of our collective consciousness, and has given a new meaning to the word *worldview.* It serves well as a needed public reminder that we are all on the same planet, and that global civilization is much more than an idealist's dream.

But while the photograph is a powerful piece of visual information, it is also misleading. It is, after all, a *still* photograph. It doesn't show how the world is changing, and that is the really interesting part—the part without which no worldview is complete.

The Earth is undergoing a sort of mutation, growing an incredible network of new information and communications systems: wires and fiber optic cables are extending and connecting, messages are bouncing off of relay towers and satellites, people are purchasing telephones and radios and computers and other electronic systems by the millions. This process—"cybernation," as Donald Michael calls it—alters the way the world works by creating new connections among ecosystems.[3] It alters human society by creating new linkages among people—entirely new communities and new *kinds* of communities unbounded by geography. And it alters the relationship between the human species and the planet by giving people new access to information about what's going on in the world. We are now able to see, indeed, *forced* to see, a world that is restless, turbulent, and inventive—vastly changed from what it once was, and in certain ways changing more rapidly now than it has in the past.

At the same time that a new genetic science is emerging, so is a new science of global ecology. Because of the way we have compartmentalized knowledge in the past, these are generally taken to be two entirely different developments—but they really aren't. Microbiology and macro-ecology are both sciences of life, they are both being transformed by the information revolution, and they are both participating in the creation of a new world.

This new world is a mobile world, in many ways. Symbols of all kinds—data, news bulletins, entertainments, ideas, cultural fads—flow about at unprecedented speed and volume. Information moves, people move, animals move, microorganisms move. Even plants move; the average home garden is a gathering of foreigners. And day and night, the satellites glide silently through the skies, playing a greater part in human life than most of us recognize.

Gaia's Gadgets

Satellites have become the real wonder-workers in present-day environmental information-gathering, and they are evolving—becoming more sophisticated, more intimately linked, and more important to the future of humanity and all life on the planet.

In 1955 the visual symbol that was adopted by the leaders of the International Geophysical Year project—the first global-scale coordinated project of Earth studies—showed the planet being encircled by an artificial satellite in orbit. The logo was pure wishful thinking. No artificial satellites had yet been launched. But the scientists behind IGY expected that satellites would be launched in time, and that eventually they would play a part in their research. They could hardly have suspected that within the next thirty-five years, 20,000 satellites would be put into orbit.[4]

Twenty thousand. Some of those have come flaming back into the atmosphere by now, of course, and others are finished with their tasks and are wandering around out there with nothing to do. But at any time thousands are busily carrying out their various missions. Many of them are environment-watchers. They track storms around the world, sample the health of vegetation, follow the migrations of wildlife, note the expansions and contractions of desert and forest, take the temperature of the oceans. Others serve as parts of the information systems that link laboratories, data bases, researchers anywhere in the world.

Never before has the world been so relentlessly mapped and monitored. Never before has such a flood of data been beamed back to keep us all informed about how well—or how badly—the biosphere's health is holding up. This is a part of how we go about becoming a different kind of species, while the Earth becomes a different sort of world.

Lewis Thomas wrote: "I have been trying to think of the Earth as a kind of organism, but it is no go. I cannot think of it this way. It is too big, too complex, with too many working parts lacking visible connections. The other night, driving through a hilly, wooded part of southern New England, I wondered about this. If not like an organism, what is it like, what is it *most* like? Then, satisfactorily for that moment, it came to me; it is *most* like a single cell."[5]

It doesn't really pay to quibble with metaphors, especially when elaborated by a celebrated biology-watcher such as Dr. Thomas. But sometimes I go into the High Sierra, and lie in my sleeping bag at night considering the stars, and as I do I note that many of them turn out to be satellites. I reflect on the meaning of those 20,000 exquisite little devices sent out to cruise silently through space, and I conclude that anything able to surround itself with such information-seeking technology would be one hell of a cell. No, the Earth we live on now is something else, something we really don't have a word for, something that beggars our best metaphors.

Satellites are only a part of a rapidly evolving environmental information technology. Wildlife management depends heavily on radio transmitters carried about by everything from migrating birds to foraging caribou, and equally fancy new tools are being used for such tasks as soil sampling and air pollution measurement. It is ironic in a way that the protection of nature should become wrapped up with the use of amazingly sophisticated electronic instruments. But that is exactly what is happening.

The Age of the Smart Map

"Computer models are revolutionizing science," reports an article in *Earth* magazine. The statement itself is hardly surprising news at this point, but the article is referring to an aspect of the modeling revolution that is often underestimated or misunderstood: its growing importance in studies of the Earth. Computer modeling by scientists on the ground is what breathes life into the data collected by satellites in orbit. Climate simulations are probably the best-known products of this technology—you see them every evening on television when your friendly meteorologist speculates about tomorrow's weather—but no less important are the models of ocean systems, and those that simulate the deep rumbles and flows beneath the planet's surface. These models are never perfect as describers or predictors (all information, as we'll note in the next chapter, is incomplete), but they can be continually revised. They are (as the same article puts it) "dynamic, electronic encyclopedias that can easily be updated as new findings from the field come in."[6]

I prefer to think of them as smart maps. The map, as the great semanticist Alfred Korzybski pointed out decades ago, is not the territory—and that will ever be so, even if someday somebody draws a map like the one described in the fantastic fable by Jorge Luis Borges, a map so large and detailed that it precisely replicated the kingdom it was meant to describe. But computer models enable us to see the territory as people have never seen it before—to peek into its life processes, watch it change, imagine its future. They do something that maps have never done before—show not only the lay of the land, but the things that are happening to it.

What's happening in this field is a lot like what's happening in medicine. It's impossible to miss the similarity between this new science of mapping and modeling the biosphere, and the science of mapping and modeling the human body. One method used to obtain data for models of the Earth's mantle is essentially a CAT scan—the difference being that seismic waves are used instead of X-rays. The waves travel through the planet, and their changing speed pinpoints hot and cold patches of rock far below the surface.

Global Citizens

It's quite a recent plot development in the storyline of human evolution that people began to have any concept of the world's total population and its rate of increase—and began to worry about it. The Rev. Thomas Malthus published his celebrated "Essay on the Principle of Population" in 1798, and population growth became a general—though not particularly high-profile—topic of discussion in the nineteenth century. Since then the population has kept growing, and so has the science of demography. Now there are institutes of demography, journals, departments in universities. Demography has become our bio-information science *par excellence.* Population figures are published in beautifully illustrated books with colorful map graphics showing which countries are the most crowded, and no corporation or public agency can do its work without population projections—usually in multiple scenarios, regularly revised to incorporate the latest data. Population organizations distribute all manner of demographically related material: today's mail from Population Action International offers me a booklet on water supplies (using population projections to estimate per capita availability of fresh water to the year 2050) and another report that ranks countries according to per capita availability of arable land. Demographic studies are the heart of macro-ecology, because the large-scale changes in the world are the products—deliberate or accidental—of human activity.

Currently, demographics is going through a transition—not so much a paradigm shift as a shift of emphasis. It has become increasingly clear to demographers, and to all the political activists and public officials and others who concern themselves with demographic information, that there are two "population problems." One is the problem of growth, the same one that has exercised people from Malthus to *Population Bomb* author Paul Ehrlich, and the other is the problem of movement.

An interesting demographic fact of our time is that we are living in the midst of the greatest mass migration of all time. More people are in motion—tourists, business people, refugees, legal and illegal immigrants—than ever before. They move within countries, from country to country, and in a bewildering variety of patterns—rural-to-urban, rural-to-rural, urban-to-urban, urban-to-rural, circular—that are almost impossible to capture on a two-dimensional map. And as they do, a profound but little-noticed change takes place in the world: in any given area or community, there are likely to be more people who have at some time in their lives been migrants than people who haven't. As I recently heard one demographer sum it up: "Migration, rather than stability in one place, is now the norm." She also pointed out that in the information era, the connections to people and place in the country of origin are more easily maintained by those who emigrate. Migrants call home, send money back to their families, return for a visit or to retire. And information technologies create entirely new kinds of linkages, such as computer hookups among former Nigerians or Brazilians now living in other parts of the world. "What is happening as a result of all types of migration," she reported, "is multi-locality. Individuals, families and communities are no longer rooted in one place, nor are they placeless. Rather, people have multiple linkages to multiple places."[7]

This is a remarkable development in itself, but we need to take it a step farther and recognize that people not only have linkages to multiple places, but also have linkages—increasingly concrete and visible ones—to the whole world. Today we are experiencing a revolution of expanding boundaries as people—usually without deliberate effort, without self-consciousness, without noticing—revise their perceptions of social and geographic space, and become parts of a larger whole. This may or may not result in actual physical movement. The mass media of news and entertainment involve people in distant events, globalize all manner of facts and fantasies. The various processes and systems of globalization—economic, cultural, political—reach out and touch people even in remote places, draw them into new relationships.

For the first time in human history, everybody is beginning to live in the whole world.

Some people are more global than others, of course, some embracing the new situation and some fiercely resisting it—but sooner or later the world finds us all. Global citizenship, once an abstract aspiration of a few high-minded idealists, is now a commonplace reality of everybody's life.

The Other Global Citizens

And people aren't the only world travelers. Life forms of all sorts and sizes are also in motion: pets, animals for zoos, tropical fish, ornamental plants, specimens for research, seeds for agriculture, frozen bull semen and frozen embryos for animal breeders, killer bees migrating northward, insects stowed away in fruit shipments, viruses and bacteria stowed away in people.

There is economic globalization, cultural globalization, and political globalization—and there is also biological globalization, the least noticed but perhaps most powerful force of all.

We are living in the midst of the greatest global movement of plants, animals, insects, and microorganisms of which there is any human record. As this happens, the ecosystems around the world, particularly the more hospitable ones, are becoming as "multicultural"—as thronged with immigrants—as the nations and the cities. And just as the science of demography evolves toward an increasing emphasis on populations in motion, the sciences of botany and wildlife biology and microbiology develop a new sensitivity toward the phenomenon of plants and animals and microorganisms that commute about the world. We are informed by facts like these:

- The nocturnal Melanesian brown snake was accidentally introduced into formerly snake-free Guam during World War II. The brown snake population on Guam is now approximately 12,000 per square mile, and eight species of native birds have become extinct as a result.[8]
- According to the U.S. Fish and Wildlife service, American box turtles—popular as pets in Europe—are being exported at the rate of 25,000 to 30,000 a year.[9]
- The state of Florida is infested with invaders including Amazon pythons, South American lizards, and Asian tiger mosquitoes. An Australian ornamental plant, *Melaleuca quinquenvervia*, crowds out native plants, forms artificial islands in the Everglades, and dries up wetlands. Its resin burns easily, feeding wildfire. The hudrilla, imported from Sri Lanka for use in aquariums, has overgrown more than 40 percent of the state's rivers and lakes.[10]

We all know about some *parts* of biological globalization. National governments and international agencies struggle constantly to regulate or control various biological wanderers. In California we worry a lot about Mediterranean fruit flies, which lay their eggs in citrus fruits. The larvae tunnel into the fruit and cause massive crop damage. Farmers, of course, regard the flies as malevolent outsiders. The fruit flies, however, feel pretty much at home here—as well they might, since the state is full of Mediterranean fruit trees. Many native trees—chestnut, elm, hemlock, red pine—have also been severely damaged by foreign insects, blights, and fungi. Various efforts are now being mounted to protect them—everything from designing genetically-engineered viruses that attack the fungi to breeding a new generation of American chestnuts with a resistance gene from a closely related Chinese species. Biological pest control, favored by environmentalists, also tends to advance biological globalization. In northern California there is a monument to a beetle, surely one of a kind. It celebrates a leaf-eating insect, imported and released after a worldwide talent search, that brought an end to the trouble caused by a European weed (St. John's-wort) that poisoned grazing sheep and cattle.

This is the world we live in, this orb of biological jet-setters. Most of us don't quite know we live in such a world—there is no photograph to give us the message in a single image. Some specific elements of the biological globalization process can be managed or controlled to some extent, but the larger process is irreversible. It is creating a different biological reality, and we are not going back to the world as it was.

Put this way, biological globalization appears to be only a problem, and an insoluble one at that. But there is another dimension to it—the globalization of biological resources in gene banks—that, although problematic in many ways, is also one of the great wonders of the emerging bio-information society.

The New Gold, the New Fort Knox

The international network of gene banks is already central to the world's food production system, and it is destined to become an incalculably precious global resource in the near future.

The ancestors of the modern gene-banking network were the pleasure-gardens of kings and emperors, often grown with rare and wonderful plants from distant lands, and the "medicinal gardens"—one of them, in China, established in 2800 B.C.—that gathered herbs known to be of value in treating human diseases. The best known and probably largest botanic garden was the Royal Botanic Garden at the London suburb of Kew, which was established in the mid-eighteenth century and now holds some 50,000 different species of plants—ornamentals, medicinals, and wild relatives of crop plants. There was always a global dimension to this kind of collecting, because the whole point of establishing such a garden was to expand the range of plants available for the use of the ruler and his subjects, and so explorers and sea captains were expected to bring back decorative or possibly useful new plants. Also, collecting specimens easily led to trading specimens, so particularly useful plants—such as rice—tended to become global citizens if they were sufficiently hardy to survive transplantation to different soils and climates.

There is much more to gene banking now than merely gathering samples, just as there is more to financial banking than storing money. The gene banks have become an insurance program against loss of genetic diversity. This diversity is rapidly disappearing in the wild as a result of human population growth and spreading agriculture, but there are gene banks all over the world. Some are large libraries of genetic information with species from many places, some specialize in a certain crop—such as rice—and some focus on the crop plants central to the agriculture of a certain country or region. Some, such as the U.S. Department of Agriculture's National Seed Storage Laboratory at Fort Collins, Colorado, are supported by national governments, others are private or university-based collections. The major ones are affiliated with the Consultative Group on International Agricultural Research, a global network established in 1971. Linked by telephones and e-mail, they exchange information around the world.

A gene bank is usually several things at once. It is, first of all, a storehouse of genetic information, which may be in the form of seeds, clippings, live plants, live animals, tissue specimens, frozen embryos, or DNA segments. It is also a storehouse of data, without which the genetic information is of limited value. A well-run gene bank contains huge files of what is called "passport information" on each of its accessions—recording where and when the sample was collected, some description of the ecosystem from which it was taken and any further scientific information on its characteristics.[11] Each sample that is stored, if it is in the form of seed or tissue, should also be grown out periodically, evaluated, tested for resistance to disease and insects. More data accumulates in the process. So the gene banks are computerized. This has not always been a smooth operation—a study in the 1980s found all kinds of inconsistencies in computer systems, and a limited ability to exchange information—but all the major gene banks are now linked and able to talk to one another.

Gene banks are also research centers. The Green Revolution came out of work done at the International Rice Research Institute in the Philippines and the wheat-and-corn think tank, Centro Internacional Para Mejoracion de Maiz y Trigo (CIMMYT), in Mexico. Normally all gene banks make seeds or clippings available to any legitimate scientist who requests them.

The good news about gene banks is that they are often extremely effective in improving food crops and helping farmers recover from disasters. The bad news is that the global system is not at all ready to assume its global responsibilities. Let's look at the good news first.

The International Crops Research Institute for the Semi-Arid Tropics (ICRISAT) specializes in the food crops of the hot, dry climates that are found in parts of many developing countries in Africa, Asia, and Latin America. These are not the world's most desirable farmlands, generally speaking, but they are the home to many people—around one-sixth of the world's population inhabits such regions—and they produce a variety of foods such as sorghum and pearl millet. In the 1970s ICRISAT and another research center in Syria began looking for a better chick-pea. Chick-peas are an enormously important food crop in these dry areas—and also a crop that has often failed. The traditional practice used to be to plant them in the spring and harvest them in the fall, but that left them vulnerable to the intense and often destructive summer heat. If a variety could be bred that could be sown in winter, it would most likely increase productivity.

So the researchers went to work searching the genetic libraries, screening thousands of samples, interbreeding currently farmed chick-pea varieties with wild relatives, finally getting down to a short list of forty-two cultivars that were then sent out to other researchers in fifteen different countries where they were grown and tested. The final selections were made available to farmers in 1988, and by 1993 some 90,000 hectares were growing winter chick-pea, with a reported increase of 60 percent in yield and 100 percent in profits to the farmers.[12]

Targeted plant-breeding efforts of this sort are information-management projects. It is true that excellent results have often been achieved in the past by solitary experimenters working in circumscribed geographic areas, but the major breakthroughs needed today are much more likely to be achieved by international research teams who are able to pool their efforts and search efficiently through vast amounts of genetic information.

The ghastly incidents in Rwanda in the mid-1990s led to another impressive performance by the gene-banking system—in this case, a genetic rescue effort.

The whole world watched the slaughter that broke out in Rwanda in 1994. William Scowcroft, an Australian plant geneticist at the Centro International de Agricultura Tropical in Columbia, saw from the beginning that it was going to be an agricultural disaster as well: a growing season was being lost, farms were being despoiled, communities were being broken up. He and other members of the global gene-banking community knew that even if some semblance of peace were restored, the knowledge base of the country's agriculture would be lost. So they set in motion a project that came to be called "Seeds of Hope." Various gene banks had information about Rwanda's main food crops. ICRISAT, the semi-arid tropics specialist, knew about sorghum. CIMMYT in Mexico knew about the maize. Centro Internacional de la Papa in Peru knew about the potatoes. The International Institute of Tropical Agriculture in Nigeria knew about the cassavas. Those were the data sources. Other gene banks, most of them in Africa, had samples of Rwanda's indigenous seeds. Planting programs were started in Burundi, Tanzania, Uganda, and Columbia—surrogate mothers, a reporter called them, for Rwanda's seeds, keeping the genetic resources in readiness so that whenever Rwanda passed through its human crisis, it would not proceed directly into an agricultural one.[13]

That was an example of what can be done in response to a disaster —whether act of God or act of man—that destroys the genetic base of a country's agriculture. But it also relates to the bad news, because it is highly doubtful that the global gene banking system is capable of responding to disasters in the future—or even of surviving disasters of its own. Although the world's political leaders have begun to discover that there are such things as genetic resources, the protection of such resources is still not really high on the political priority list.

Gene banks vary tremendously in size and quality. The one in Fort Collins, Colorado, has the world's largest cold-storage vaults for seeds, but is not considered the best managed; that honor goes either to the rice-research institute (IRRI) in the Philippines or to the wheat and maize center (CIMMYT) in Mexico. Some gene banks are small, and consist of little more than a few seeds and a few plants.

The data bases are seriously inadequate. One report says: "Of the world's germplasm collections, some 65 percent have no passport information and between 80 and 95 percent lack characterization or evaluation data. . . . Even at well-managed gene banks, many potentially useful genes remain unused because evaluation has lagged behind the pace of collection. Three-quarters of IRRI's rice collection has been evaluated, but only 10 percent of the potato accessions at the Centro Internacional de la Papa have been thoroughly tested for their characteristics."[14]

Then there is the matter of the seeds themselves. In the best of the long-term storage units, seeds are partially dried and stored at subzero temperatures, where they may remain vital for up to half a century. But sooner or later, the seeds must be grown out and the stores renewed. And many of the seed storage facilities are far from ideal. Major M. Goodman, professor of crop science at North Carolina State University, summed up the concern of many scientists a few years ago when he said that "a germplasm system which merely acquires material and does not have facilities for evaluation and utilization is really not a system at all." He identified collections in the United States and around the world among these nonsystems and added: "I maintain that seed banks holding such collections are really seed morgues. What goes in is not going to come out alive."[15]

Gene banks are meant to be more secure than the environments from which the samples were taken, but they are vulnerable in many ways. Seeds can mold or be eaten by rats or insects. Cold-storage vaults can be hit by power failures. Growing crop plants and trees are subject to fires and storms, and to the kind of disaster that struck a gene bank in Somalia when starving people broke into it and ate much of its collection. So, in order for the gene-banking system to function as a global resource—able to respond to large-scale or small-scale disasters, able to contribute to the massive agricultural response that will be necessary in the event of global warming—individual gene banks and the system as a whole must be improved in many ways. It must, among other things, have backup germplasm collections and data banks so that a local mishap does not become an unrecoverable global loss.

So, much concern and controversy surrounds the gene banks: about the system's adequacy and safety, about whether the money-poor but gene-rich developing countries from which samples are taken should be more adequately compensated, about the generally low level of interest in the whole subject among political leaders and the public, about how germplasm should be best protected. Conservationists are in favor of a much stronger emphasis on *in situ* preservation—that is, protection of genetic diversity in the field, and protection of ecosystems—over exclusive preoccupation with *ex situ* preservation in gene banks. Peter Raven, director of the Missouri Botanical Garden, says: "What might be called a Noah's ark strategy is fine, but it neglects the need for managing plant and other communities throughout the world in a sustainable way, so that the cryogenically stored seeds, or whatever other remnants of wild populations we may have saved, can be used someday. Otherwise, happy with the seed banks we created in our basements, we could probably get on with our lives while, metaphorically, Rome burns."[16]

I have been mainly talking about gene-banking of plants, and about the essentially agricultural problem of "genetic erosion"—loss of diversity within species—that results as ecosystems change, or as farmers abandon local varieties in favor of higher-yielding new breeds. But we also need to be aware of the critical importance of gene banking to protect against the loss of entire species, and of the loss of species and variations among animals and even microorganisms.

The same forces that have caused the homogenization of the world's agriculture—in which farmers have been wooed away from the old "land races" of locally developed plants—have also caused people to abandon their traditional breeds of farm animals. As noted in a publication of the UN's Food and Agriculture Organization (FAO): "The genetic diversity among animals deserves to be preserved for future generations as much as do other manifestations of our cultural heritage, such as art or architecture. Developed over hundreds, perhaps thousands of years by breeders responding to the changing fortunes of history, domestic breeds are, as much as any work in paint or stone, the fruit of human genius."[17] And so we have a growing global system of animal gene banks, *in situ* and *ex situ*, with corresponding data bases. The FAO is developing a Global Program for Animal Genetic Resources, with regional animal gene banks, a global inventory of animal genetic resources, and a Global Animal Genetic Data Bank located in Hannover, Germany.

There is some similarity to the plant gene banks: cryogenic storage of semen and embryos is now possible, and germplasm in that form can be stored for several decades. A pioneer in this kind of gene banking was the famous "frozen zoo" in San Diego—a research annex to the zoo itself—which contains not only frozen sperm and embryos, but also frozen tissue samples from various rare animals in the zoo. In the future it may be possible to reconstitute an entire animal, a la *Jurassic Park,* from a few cells. (It may also, some scientists believe, be possible fairly soon to reconstitute a simpler organism such as a bacterium from no cells at all—only the information on its genome stored in a computer's memory.) In the meanwhile, it is possible now to implant an embryo from, say, a rare and endangered type of African deer into some distant relative such as an American mule deer. Rare domestic sheep, goats, oxen, and other animals are preserved in agricultural theme parks such as Plymouth Plantation in the United States and Cotswold Farm in the United Kingdom, and also raised by hobby farmers and other private breeders. In Canada the province of Quebec subsidizes farmers who rear purebred Canadienne cattle.[18]

The role of zoos is going through a metamorphosis as the global gene banking system evolves. The environments of zoos more closely resemble natural habitats now; zoos become homes for endangered species and sometimes support restoration activities through their captive breeding programs. They also become research centers working in cooperation with *in situ* conservation efforts. If you dropped in on the curator of mammals at the Bronx Zoo, you might find him sitting in front of a computer screen watching a displayed map of Africa that shows—with the help of an orbiting satellite—the movements of forest elephants in the Cameroon. The elephants are fitted with radio transmitters. Using the data broadcast from them and received by the satellite, ecologists in the Bronx follow the herds' movements and behavior patterns—much more closely, in a way, than wildlife biologists were ever able to do in the dense jungles—to work out conservation plans. Conservation plans for wild animals are information-intensive activities, and the zoos—by providing information on the nutritional needs and breeding habits of animals, and testing radio transmitters—create a synergy with *in situ* conservation efforts.

Finally, bacteria. Most of us can't get too enthusiastic about saving germs, but an international group of microbiologists recently drafted an appeal for help in creating a system—capturing samples, storing them, and gathering data—to preserve bacteria living in rare (and perhaps endangered) environments such as hot springs and hypersaline lakes. The document pointed out the possible utilitarian payoffs, noting that the bacteria in hot springs contain enzymes that work faster than those of normal bacteria, and that the salt-tolerant types might be useful in a variety of new scientific and technical applications—another gene bank.

The gene-banking system leaves a lot to be desired, and yet it—I speak of it now as one system, which it must gradually become, however loosely linked its various components—is a quite impressive and amazing development: bionics on a global scale. Meanwhile, systematic biologists—the people who study biological diversity—are urgently advocating another "big science" project. They call it Systematics Agenda 2000. If it comes to pass, it will be a worldwide inventory of all living things, a global analog to the Human Genome Project. More information gathering, Aristotle's project carried into the twenty-first century.

Biosphere Three and the Electronic Noosphere

Pierre Teilhard de Chardin, the theologian who was also a paleontologist, likened the stages of the Earth's evolution over time to the layers of its zonal composition: the central and metallic barysphere surrounded by the rocky lithosphere, in turn surrounded by the fluid layers of the hydrosphere and the atmosphere. The emergent stages that most interested him were the formation of the biosphere—"the living membrane composed of the fauna and flora of the globe"—and the noosphere, the realm of conscious thought that grew out of the biosphere and transformed it. "Much more coherent and just as extensive as any preceding layer, it is really a new layer, the 'thinking layer,' which, since its germination at the end of the Tertiary period, has spread over and above the world of plants and animals. In other words, outside and above the biosphere there is the noosphere."[19]

The term *biosphere* has come into common usage now—not always employed quite the way Teilhard and his co-inventors of the term intended it, but approximately synonymous with Earth itself. A good and serviceable word, I think, because it reminds us that the planet is alive; somehow, when I look at that photograph of the Earth, I find it feels right to call it a biosphere. That other coined word, *noosphere,* (from the Greek word for "mind") scares some people off, seems to border on the flaky. And yet not long ago I heard a roomful of scientists using it shamelessly, nodding their heads in agreement at its appropriateness. This was at one of the annual gatherings of the American Association for the Advancement of Science. AAAS meetings are always interesting because they bring together world-class scientists from different disciplines who are required by the nature of the event to speak across disciplinary boundaries, and thus to employ something more or less resembling the English language. This particular conference featured a number of presentations on the subjects we have been discussing here. The scientists talked about enormous research undertakings, each in itself greater in scope and complexity than any yet undertaken by the human species. They talked about astonishing new uses of computers and information technology. And in one of those sessions the participants, with an unconcealed sense of wonder, talked about how all those things *come together*—are happening in the same world, are parts of a single process. And "noosphere" was the only word that seemed to do it justice.

I don't think Teilhard de Chardin had any concept of the noosphere as an entity; it was more an invisible ocean of thought that floated about the Earth. But now the noosphere is growing its own electronic organs, connecting the whole world and all its people in very tangible ways. The "thinking layer" of the biosphere has become much less an abstraction: it is now physically embodied in the satellites that hover above the planet, observing all its life processes and sending us information.

Another item from my notebook: some years ago I had an opportunity to take a tour of Biosphere II, the enclosed artificial ecosystem down in the Arizona desert. This was just before the beginning of the much-publicized experiment in which a team of "biospherans" were sealed into it for a two-year test run, before the publicity barrage and the heavy criticism of the project as flawed, if not downright foolish, science. Not feeling any particular need to pass judgment on Biosphere II, we simply enjoyed it—enjoyed wandering through it and surveying the diversity of life forms and mini-ecosystems it contained, enjoyed looking at it from outside and appreciating the beauty of that huge structure as it squatted like some bizarre, silvery spaceship amid the dry hills.

I particularly remember its nerve center. It was a spacious, modernistic room with many monitor screens mounted on the walls, and it reminded me of the bridge in the *Starship Enterprise*. It was a place into which information flowed from all parts of the Biosphere, telling at any moment the condition of its various parts and subsystems. And the information did flow in, because the whole Biosphere was wired. And it struck me that, although Biosphere II was meant to be a small and simple model of Biosphere I—the Earth—the real planet is more and more coming to resemble the model. Earth, too, is getting wired together in a new way, information is flowing back to our various monitors. There isn't a single nerve center—I hope and suspect there never will be—and nobody can simply control all the global systems of life, push a few buttons and change the weather or the undersea currents. But people do affect all those systems, and the electronic noosphere is becoming much more adept at showing what happens when we do.

CHAPTER FOUR

Welcome to the Bio-information... Society

The growth of the external memory system has now so far outpaced biological memory that it is no exaggeration to say that we are permanently wedded to our great invention, in a cognitive symbiosis unique in nature.

—Merlin Donald[1]

For a couple of decades now various prophets and gurus have been telling us about an event called the information revolution, which, they say, is bringing in a new order of things called the information society. Most of the heralds are enthusiastically in favor of this development, a few are plunged into pessimism about it—but they agree that it is a major world-historical happening. They tell us that our increasing dependence on information and our increasing use of new information and communications technologies are changing politics, changing governance, changing business, rewriting the rules of organizations everywhere.

Daniel Bell, one of the first of the information theorists, called information "the strategic resource and transforming agent of the post-industrial society," the central pivot in "a new social framework based on telecommunications." In his view, the three key features of the post-industrial information society are: (1) the shift from a goods-producing to a service economy, (2) the increasing reliance on theoretical knowledge, and (3) the creation of a new "intellectual technology" based on computers and other smart machines.[2]

Peter Drucker, another information pioneer, talks about how information creates more flexible, less hierarchical organizational structures and sets the stage for a new class of "knowledge workers" who require different styles of management. He says we are moving into a "post-capitalist" era in which information replaces capital as the generator of wealth. Drucker's favorite economist, Paul Romer, says that information and information technology are capable of bringing about a permanent change in the rate of discovery and the rate of economic growth. He believes that the world is now poised on the edge of an unprecedented burst of innovation and wealth creation as people break free of traditional economic thinking and industrial-age limitations.[3]

Harlan Cleveland talks about the "informatization" of world affairs and says the concept of resources that went with traditional "geopolitics" is rapidly becoming obsolete.[4] "Nowadays," he says, "it's the countries with the biggest flows of information we call 'developed.' We know that anybody can extract knowledge from the bath of information that nearly drowns us all. You don't have to find it inside your own frontiers, you don't have to grow it in your own soil, you don't have to fabricate it in your own factories or put it together in your own assembly plants. You do have to 'get it all together' in your own brain, and then combine your insight and imagination with other human brainwork in networks, companies, alliances."[5]

Donald Michael, another of the pioneer information theorists, stresses the importance of *learning* as societies become increasingly dependent on the creation and exchange of knowledge. In this he is not merely making the standard endorsement of education, but arguing that a whole new set of learning skills—which he calls the "new competence"—become necessary for anyone who aspires to function effectively, and especially for people in positions of leadership. Leaders have to be "learning leaders" who can acknowledge error and revise their plans, and organizations—including entire societies—have to be able to shift into "learning modes" that may produce fundamental changes in their ways of doing things.[6]

My Meridian Institute colleague Steve Rosell produced a book entitled *Governing in an Information Society*—a report on a project in which he and a group of high-level officials of the Canadian government explored the impact the new order of things was having on their own society and political system—and it noted the following main effects:

- The trend to globalization: including the globalization of the economy (interconnected stock exchanges, frontierless capital markets, globalization of manufacturing and so on), the pervasive influence of globalized science and technology, and the growing need to handle issues, from trade to the environment to human rights and more, in supranational forms, networks, and organizations.
- *Atomization, democratization, and fragmentation:* reflected in the increasing power of subnational governments, in growing regionalism,

and in the proliferation of "multiple voices," that is, the increase in the number of groups organizing to assert a role in the process of government.

- *A breakdown of the bureaucratic/industrial model of organizing:* both public and private sectors are downsizing, stripping away middle management, contracting out work, and relying more on networks and task forces and other more flexible, decentralized ways of organizing.
- *A fundamental restructuring:* the breakdown of the historical distinctions between industries, between the public and private sectors, and even between states, accompanied by a search for new relationships and alliances between those entities.
- *The decreasing possibility of secrecy,* and the implications of that for governing systems that rely on a certain degree of confidentiality.[7]

Those are some of the prevailing general ideas about life in the information society. The specifics include NAFTA, GATT, the Internet, virtual reality, the collapse of the Berlin Wall, the implosion of the Soviet Union, the increasing feistiness of nongovernmental organizations, the growth of the computerized global financial system, Cambodians watching Mexican soap operas, Americans watching the war in Iraq, the whole world watching the spectacle of the moment, and whatever news story about a famous person's sex life you read this week.

Most of the above has to do with the political, economic, organizational side of the information revolution. But there's another side, no less striking and no less potent as a transformer of the world, that has to do with the connection between information technology and our personal lives. We need to understand this connection if we are to understand how the increasing emphasis on *biological* information is going to affect us. The people who are exploring this side of the information revolution see it as also an evolution, a new chapter in the development of *Homo sapiens*, an event that changes not only society and business but the fundamental nature of human life.

Howard Rheingold—author of several books on cognitive science, virtual reality, and related subjects—says the real visionaries in the computer world have always understood that the technology is not meant merely to store or process data, but "to extend the power of human minds to think, communicate, and solve problems."[8] For them, the term that best describes what computers do is *augmentation*. That word is the key to appreciating the uniqueness of human evolution—its past and present, and especially its future.

The Augmented Animal

A team of archaeologists at Indiana University's Center for Research into the Anthropological Foundations of Technology have been studying the simple implements of flaked stone—called "Oldowan tools"—that were used by early hominids some $2\frac{1}{2}$ million years ago for purposes such as slicing through the inch-thick hide of an elephant. These tools, they say, were an integral part of the hominids' development as carnivores, evidence that they had begun to "produce simulated biological organs—slashing, crushing organs" like those of other carnivores. "You see a reduction in the size of the teeth and jaws because we're replacing biology through technology," one of them explained. The tools were early augmentations that freed the body to evolve in different ways.[9]

Marshall McLuhan, whom we remember chiefly for his provocative ideas about television, had a good insight into this. His book *Understanding Media* was subtitled *The Extensions of Man,* and in it he pointed out that all kinds of inventions have an augmentation effect. They serve as extensions of the human body and enable people to do things they can't do with the physical equipment they had developed through genetic evolution. He even saw clothing as an extension of the human skin—and a most effective one, since it conserves bodily energy and enables people to survive with less caloric intake. He went on:

> If clothing is an extension of our private skins to store and channel our own heat and energy, housing is a collective means of achieving the same thing for the family or the group. Housing as shelter is an extension of our bodily heat-control mechanisms—a collective skin or garment. Cities are an even further extension of bodily organs to accommodate the needs of larger groups. Many readers are familiar with the way in which James Joyce organized *Ulysses* by assigning the various city forms of walls, streets, civic buildings and media to the various bodily organs.[10]

Augmentations—artificial additions to or improvements on the physical organism—are by no means exclusive to human beings. Evolutionary biologist Richard Dawkins uses the term "extended phenotype" to describe all the tools and structures that animals create instinctively.[11] But human augmentations become parts of culture—inventions and ideas that can be passed around, improved, and used in new ways. And there's an even greater difference: human beings also augment *thinking.*

Psychologist Merlin Donald of Queen's University in Ontario has brought together findings from a number of fields into a masterful theoretical framework for understanding what he calls the cognitive evolution of the human species.[12] His approach also might be called evolution by augmentation: augmentation not of the body, but of the capacity to comprehend and communicate information.

His thesis is that the human species pulled itself upward from the primate level by a series of inventions. This in itself is not an entirely new concept; it recollects Bergson's *Homo faber*[13] and other ideas of a human as the tool-making animal. But Donald is trying to make a larger and subtler point. He says that the most important of these inventions were systems of representation and cognition. People first invented symbols, things that stood for other things—a remarkable invention in itself—and then proceeded, over long periods of time, to invent entirely new, ever more complex and powerful symbolic systems such as speech and writing. And each of these new systems of representation marked an evolutionary transition to a new level with new capabilities of thought and action—in effect, produced a new species that was fundamentally different from the one that had preceded the invention. Each was an evolutionary transition, and each changed not only humanity but the world.

In the beginning was not the word, however. The first transition —the move from the level of apes and australopithecines to the level of *Homo erectus*—was connected with the emergence of the ability to mime, or reenact events. The transition, which itself probably took several hundred thousand years to complete, was under way about 2 million years ago. And it enabled our *erectus* ancestors to do things no previous animal had done—organize their societies in new ways, produce highly refined tools, master fire, and migrate far beyond their original habitat.

The next transition, the move from the culture of *erectus* to that of *Homo sapiens,* was accompanied by the beginnings of true human speech. And this of course involved a whole lot more than making meaningful noises. It involved developing a new cognitive capacity. People had to be able to think and talk in narrative form, and to comprehend the narratives of others. As they did they gained new powers of learning and group memory, developed much more complex tools, and created elaborate oral cultures of myth, ritual, and religion.

These evolutionary transitions were both cultural and biological. The new symbolic systems made it possible to build up bodies of knowledge that could be handed down from generation to generation— "cultural DNA," some evolutionary theorists call it.[14] At the same time, human beings evolved physically. Biological selection favored the individuals with brains fit for using symbolic information, and with vocal tracts fit for speech.

The third transition is the one Donald calls the move to "visuographic" systems of representation that used markings or drawings to stand for other things or concepts.

It seems so "natural" to us to represent the world with visual symbols that it is extremely difficult to get our minds back to a time before that development—to imagine human life any other way. But visual representation was a true invention, a breakthrough that some people made in different ways at different times and places—and many primitive societies never made at all. Archaeologists have found relics of early visuographics in various kinds of ceremonial decorations, in cave paintings, and of course in early writing. But Donald emphasizes that spoken languages don't inevitably proceed to become written ones. Out of the thousands, probably hundreds of thousands, of spoken languages that human beings have developed, fewer than one in ten ever evolved indigenous written forms, and very few of those produced significant bodies of literature—cultures based on visual representation. "Writing," he concludes, "was not only a late development, it was a very rare one."[15]

With writing came a new kind of augmentation, closer to what we would now call technology. Human thought itself began to use a tool. Carved tablets, papyrus scrolls, and eventually books became extensions of the mind, as clothing is an extension of the skin, and individuals developed a different kind of linkage to their cultures, because cultural DNA was transmitted through new channels:

> The third transition . . . led to a third stage of cognitive evolution, marked by the emergence of visual symbolism and external memory as major factors in cognitive architecture. External symbolic storage must be regarded as a *hardware* change in human cognitive structure, albeit a nonbiological hardware change. Its consequences for the cognitive architecture of humans was similar to the consequence of providing the CPU of a computer with an external storage device, or more accurately, with a link to a network.[16]

The literate human being now has an extended memory system, giving access to vast amounts of information beyond personal knowledge. And we take this for granted. Nobody expects a cook or a doctor or a historian to memorize everything connected with his or her line of work—although this is precisely what people in nonvisuographic societies had to do. In *Roots* Alex Haley recounts the experience—one of the major events of his life, and the one that made the book possible—of connecting with the remnants of such a society. In search of his roots, trying to check out a family legend of an ancestor who had been captured and sold into slavery almost two hundred years earlier, he traveled to Gambia and talked with a group of men who knew something of the country's cultural life:

> Then they told me something of which I'd never have dreamed: of very old men, called *griots,* still to be found in the older backcountry villages, men who were in effect living, walking archives of oral history. A senior griot would be a man usually in his late sixties or early seventies; below him would be progressively younger griots—and apprenticing boys, so a boy would be exposed to those griots' particular line of narrative for forty or fifty years before he could qualify as a senior griot, who told on special occasions the centuries-old histories of villages, of clans, of families, of great heroes. Throughout the whole of black Africa such oral chronicles had been handed down since the time of the ancient forefathers, I was informed, and there were certain legendary griots who could narrate facets of African history literally for as long as three days without ever repeating themselves.
>
> Seeing how astounded I was, these Gambian men reminded me that every living person ancestrally goes back to some time and some place where no writing existed; and then human memories and mouths and ears were the only ways those human beings could store and relay information.[17]

Some time later Haley returned to Africa and, accompanied by an entourage of guides, interpreters, and musicians—because he had been told that a griot did not work without background music—he found his way to a remote village. He located an old man who, in a two-hour narrative of begats, told him his family history and verified the story of his ancestor Kunta Kinte.

Nowadays, people who want to find out about their biological ancestors usually plug into the external storage systems. They buy books on genealogy, get one of those software programs with spaces for filling in the family tree, hunt up old written records, perhaps check into the huge electronic data banks such as the one maintained by the Church of Latter-Day Saints. Different kinds of linkages are made, different kinds of symbols exchanged. And far greater numbers of people connect to these networks of information. Even some of the information that Alex Haley received directly from an oral culture is now recorded in his book. People can find a copy of *Roots* in their library, or they can obtain a videocassette of the television series of some years back. Exploded through information-age media, Haley's discovery reached millions. Something is lost—we no longer train such heroic memories—and something is gained. All of literature is gained, and so is history as we know it: a history that includes more than the record of our own tribe, that spans many societies and includes human evolution itself.

The old oral cultures are vanishing, and as they do researchers scour the world with their tape recorders and sometimes with video equipment as well—talking to old men and old women, chiefs and shamans and griots, getting their stories, "preserving" the heritage of spoken lore (of course, an oral culture that is preserved electronically isn't precisely an oral culture anymore). The lore recorded is no longer confined to its original geographic and social limits. It has vaulted into cyberspace, become a part of the information society.

The evolution of culture—which preserves traces of past symbolic systems as it invents new ones—parallels the evolution of the human mind, and our past is still alive within us. "Each successive new representational system," Donald believes, "has remained intact within our current mental architecture, so that the modern mind is a mosaic structure of cognitive vestiges from earlier stages of human emergence." We develop ever more sophisticated symbolic systems. We move beyond the merely visuographic to information encoded and transmitted electronically as bits—symbols of symbols that can be translated back into words or pictures or sounds. But as we do, we also retain a repertoire of styles that enables us to communicate by mimesis and story as well.

If you follow this line of thought, you are led inevitably to the conclusion that the old puzzle about whether computers will someday be able to think better than people—the stuff of many graduate-school bull sessions, much newspaper punditry, and some serious argumentation among philosophers—is not quite the right question.[18] To paraphrase Pogo, we have met the megacomputer and he is us—us *and* an ever-changing technology in symbiosis. And as we assimilate new inventions into our working, playing, and thinking lives, we reinvent ourselves. This is human evolution, something far more than biological evolution and also more than what is usually meant by cultural evolution. As

Donald puts it: "Our genes may be largely identical to those of a chimp or gorilla, but our cognitive architecture is not. And having reached a critical point in our cognitive evolution, we are symbol-using, networked creatures, unlike any that went before us."[19]

This way of looking at the human condition, and at human evolution, is emerging in the work of visionaries from various disciplines. Historian Bruce Mazlish believes that "humans are on the threshold of decisively breaking past the discontinuity between themselves and machines," discovering "that tools and machines are inseparable from evolving human nature."[20] He calls this breakthrough the "fourth discontinuity," but what he really means is the *end* of the fourth discontinuity. His account of human evolutionary progress is a story of lessons learned—lessons of a very specific kind: with each lesson the human species discovers that things once taken to be separate are not that at all. A conceptual gap closes, a discontinuity disappears. We learned the Copernican lesson that our planet is not discontinuous from the heavenly bodies, we learned the Darwinian lesson that humans are not discontinuous from the animals, and we learned the Freudian lesson that the conscious mind is not discontinuous from its preconscious origins.[21] Now we learn that we do not simply use our tools—nor, as the technophobes would have it, do our tools simply use us—but that we and our tools together are the human organism.

Words like *tool* and *machine* are only barely adequate to describe what is happening now. We are indeed dealing with information tools and information machines, but we are also dealing with many new patterns of connection—connections between people and machines, yes, but also connections among people, among ecosystems, among all living things. And as we have seen in the foregoing chapters, much of the information that people are now using is biological. The data banks contain not only genealogy, but genetics and geology. The symbol system is becoming a tool for modifying species, modifying our own genome, modifying ecosystems. As it does, evolution itself changes. The evolution of species is no longer merely a matter of adapting to the environment. In many ways we now adapt the environment to us. Furthermore, something odd has happened to the two distinct kinds of information—genetic/biological and symbolic/cultural—that we once understood to be involved in human evolution. They seem to be flowing together, as yet another gap closes. It begins to look like discontinuities are wearing out all over the place, a whole framework of old concepts and distinctions collapsing as we struggle to make sense of what it is that we—and the world—have become. One writer in this vein, Gregory Stock, has coined another new term, "metaman," to describe the entire human species and all its far-flung creations as a single living whole. "This name," he says, "both acknowledges humanity's key role in the entity's formation and stresses that, though human centered, it is more than just humanity. Metaman is also the crops, livestock machines, buildings, communications transmissions, and other nonhuman elements and structures that are part of the human enterprise."[22]

Good Lord: bionic convergence, augmentation, cognitive evolution, the fourth discontinuity, and now metaman. The reader may well feel daunted by this feast of neologisms and wonder why writers have to keep inventing new words and phrases. Can't they make do with the vocabulary we already have? I don't think so. We are talking about a new world, and it needs some new words to describe it. Words are our main carriers of information, and sometimes we have to expand our vocabularies in order to expand our vision.

This brings us back to the information society—another recently invented term. This chapter began with a discussion of that subject and briefly reviewed some of the best thinking about it. All the principles quoted there remain valid in the context of an emerging bio-information society. In fact, if you look again at the various points raised by the information-revolution theorists, you can see their relevance to what happens when the biotechnologies and the information/communications technologies converge. The formation of a new social framework, the growing importance of knowledge workers, the informatization or cybernation of society, the changing concepts of resources, the salience of learning, the escalating globalization—all these appear to be dimensions of the next evolutionary step that is being taken, the new world that is being created, as the human species augments itself with microbiology and a global network of ecological information. It is supremely important that we understand this step as we take it, and this world as we make it.

That's what we'll be trying to do in the remainder of this book. Before we proceed into specifics, we might do well to consider some basic guidelines about information in general that are likely to prove particularly relevant to life in a bio-information society. These basic points—which I have distilled from the work of various information theorists and to which I have added a few observations of my own—will arise again in connection with many of the bio-information developments we will be discussing:

- *All information is incomplete.* There is always more to know, always another way to reframe what is already known. It follows from this that people—whether they be mothers considering an abortion or governmental leaders considering what to do about global climate change—must often make important decisions on the basis of incomplete information.

- *Information does not narrow the range of choices; it widens it.* People confronted with a difficult decision often seek more information in the hope that what they discover will, in effect, make their decision for them. They may succeed in ruling out some alternatives this way but, although further information is likely to make any decision-making process more meaningful and effective, it doesn't always make it easier. A couple that starts researching the possible ways of either preventing a pregnancy or starting one is going to find that, in the bio-information era, they have an astonishing range of choices in both directions.
- *Information is always subject to multiple interpretations and constructions.* Data is nothing until it is given meaning, assembled into narrative—we are still story-telling animals—and any piece of information can be fit into a different narrative with a different moral message, a different cast of heroes and villains. "Anything," philosopher Richard Rorty reminds us, "can be made to look bad or good by being redescribed."[23]
- *Information comes in many forms:* data, stories, myths, visual images, metatheories. Most information theorists, in fact, don't regard data as information at all; it is more in the category of what you might call potential information. Drucker says: "Data is not information. Information is data endowed with relevance and purpose."[24] Cleveland, in a similar vein, distinguishes between data (undigested facts), information (facts that have been organized for you by somebody else but not yet absorbed into your own thinking— yet-unread newspapers, for example), knowledge (information you have internalized), and wisdom (knowledge integrated, "made super-useful by theory").[25]
- *Different people speak different information languages even when they are speaking the same language.* Public controversies frequently pit people who are talking statistics against people who are talking myth—and of course they don't communicate very well. This happens frequently in bio-information controversies as the rationalists with their hard disks full of economic or scientific information bump up against invocations of Frankenstein and Gaia.
- *Information leaks.* One of the most striking—and, to many people, disconcerting—features of the information society is that nobody seems to be able to keep anything secret. A few decades ago most of the American public was shielded from the knowledge that

President Franklin D. Roosevelt was handicapped. Today, with the media gaily broadcasting to the world the intimate details of the lives of the rich and famous, such dissimulation would not be possible. This erosion of confidentiality is already becoming a matter of concern in regard to bio-information such as genetic-screening results. Information about a person's genetic susceptibility to certain environmental hazards, for example, can be used to deny employment or insurance coverage.

- *Information, once disseminated, is nearly impossible to destroy.* This final point will be a tough one for anybody who aspires to outlaw biotechnology. However attractive that might appear, it can't be done. You would have to purge every laboratory in the world, operate on the brain of every scientist and student, tear up the Internet, and burn books until you lit up the skies—indeed, repeal the whole information revolution. And still biotechnology would not go away; information has its own survival skills.

The Shape of Bio-information

The bio-information society is created by two simultaneous processes: an increase in biological information—such as the kind that is being created by the Human Genome Project and the International Geosphere-Biosphere Project—and an explosive growth of information and communications systems. These are growing rapidly now, establishing new linkages, extending and merging like the neural networks that grow in the brain of a child. In one way or another, as we will see, all living things on Earth, all ecosystems, geological, and meteorological systems are becoming connected to the new bio-information systems.

In our daily lives we move in and out of them in many ways that may seem prosaic: getting a vaccination, using a home diagnostic device, watching the weather report on television, purchasing seeds of new plant varieties for the home garden, getting or giving a blood donation, signing a form to allow our organs to be given to someone else after our death. Each of these links us to a bio-information system, each of the systems is changing, and most of these actions that we do routinely and take for granted—that we don't really notice—would have utterly blown the minds of people who lived just a few generations ago.

The bio-information society is made up of many such systems, and of all the linkages among them. These systems are not merely neutral pipelines of data. They are also, as we'll see in the following chapters, distributors of power and wealth. The information they contain is often literally of life and death importance. Furthermore, the information is often disputed. Information organizations such as corporate public-relations firms and consumer advocates sometimes go to war against one another in society, just as biological systems sometimes go to war within our bodies.

This ever-changing bio-information surround is the environment we live in now, and survival henceforth becomes a matter of learning how to adapt to it—while learning how to adapt it to us, because it is after all a human creation. We are in it, and it is in us, and for all practical purposes we *are* it. This may well be the most difficult part of the present change to understand, or—although it is quite obvious in a way—even to notice.

PART TWO

THE CHANGING LIFE OF THE HUMAN BODY

CHAPTER FIVE

Augmentations Old and New

I have about two thirds or three fourths of an erection. Occasionally it can cause a problem with bending, buckling, or slipping, so its success would depend on how cooperative a man's partner is. It's not something a bed hopper would want. Were it not for my sphincter implant, I would have chosen the inflatable type, but under the circumstances I can recommend the Flexi-Rod very highly.

—Irv, a prosthesis patient[1]

As human beings—augmented animals—we live in symbiosis with our information and our inventions. When new information and new inventions enter our lives, we change, and become different kinds of animals. And after we have changed, we rarely appreciate how different we have become from what we once were. The turmoil that attended the time of transition is forgotten, the new information fades into the background of life, the inventions are taken for granted. Most of us don't spend a whole lot of time marveling at the vaccines in our veins, the eyeglasses on our noses, or the fillings in our teeth. Perhaps we could move more effectively into the revolutionary new world of the present and future if we paid more attention to revolutions past.

Dr. Jenner's Deception

Consider, for example, immunization. It is a remarkable thing—isn't it?—that we now routinely have our immune systems remodeled, using vaccines to trick the body into developing resistance against diseases it has never had. Like many modern inventions, vaccines had a long history. Thousands of years ago Chinese doctors knew how to inoculate people against smallpox by giving them a light case of it. One way to do this was to dry and pulverize the crust of scabs taken from victims and blow the powder thus produced up the patient's nose through a hollowed bone tube. Another way was to insert an infected swab of cotton into the patient's nostril. Various skin-inoculation techniques—such as taking material from a smallpox pustule and introducing it into a small wound—were common in many parts of the world. A favorite Russian method was to cook the patients for a while in a sauna or steam bath so their pores would open up, and then to smear smallpox pus on them and vigorously pound this into the skin with birch-twig brooms.[2]

Although the practice of inoculation against smallpox or variola was widespread—in England in the late eighteenth century there were thriving "variolation" clinics—it had many shortcomings. One was that even a mild case of smallpox, with the attendant fever and nausea, was painful to endure—and was made considerably more so by the strenuous regime of bleeding and purging that doctors believed helped the inoculation "take." Another was that the case of smallpox induced by the inoculation was sometimes serious enough to cause scarring, blindness, or even death. A third was that sometimes, in the process of obtaining smallpox pus from a victim, the doctors took cultures of other diseases—such as hepatitis, leprosy, syphilis, or tuberculosis—which were passed on to the patients at no extra charge. There was no clear understanding of the mechanism of contagion, or even a consensus that the disease passed between people at all. Many doctors favored some version of the "germ" theory, which had been around for centuries, but others insisted the idea of interpersonal contagion was pure superstition and that sudden outbreaks of disease were caused by a "miasma," a malevolent, Edgar Allen Poe-ish fog arising from corpses or decomposing matter in the soil.[3]

You might think that Edward Jenner's discovery in 1798 of vaccination—that one could induce a case of relatively harmless cowpox and thereby trick the immune system into developing resistance to smallpox—would have been hailed by one and all. Smallpox, which had been introduced into Europe by returning Crusaders, was a major blight that killed people—mostly children—by the millions, and left millions more blind, maimed, and disfigured. But revolutions don't occur so easily, and Jenner's discovery of a rare instance in which immunization against one disease confers immunity to another was revolutionary in two ways: it changed medicine—it was really the beginning of immunology as a science—and it changed government, because some people began to advocate public programs of universal vaccination. As one historian puts it, Jenner was the first person in history to transform a clinical disease into a wholly social or societal disorder: "He turned smallpox into the first major disease to be made completely preventable by massive societal intervention."[4]

Partly because of this political dimension, vaccination became a hotly controversial subject. It was opposed, for fairly obvious and not too praiseworthy reasons, by some proprietors of variolation clinics, who correctly foresaw their businesses being wiped out as people switched to a method that was easier and cheaper and did not require hospitalization. It was opposed by Thomas Malthus, the original too-many-people theorist, who—correctly—foresaw that it would remove a control on population growth. It was opposed by clergymen who believed that it was an unnatural act, contrary to God's will, to give human beings a disease from cows. It was opposed by scientists and medical people of the miasma persuasion who refused to accept the germ theory. The great nurse Florence Nightingale insisted to her dying day that smallpox "grew up" on its own volition in unsanitary conditions, without being caused by anything from outside.[5] It was opposed, after the publication of *On The Origin of Species,* by Darwinists (not including Charles Darwin) who believed it subverted the healthy laws of natural selection by assisting the weak to stay alive and possibly reproduce. Herbert Spencer, the fierce Darwinist who had coined the phrase "survival of the fittest," dedicated a large part of his life to fighting against any measures that might lead to what he called "the artificial preservation of those least able to take care of themselves."[6] It was opposed, also, by people who feared that it might have some horrible side-effects; the technophobia of the time was vividly expressed by the cartoonist James Gillray, whose illustration entitled "The wonderful effects of the new inoculation" showed people with miniature cows bursting through their skin.

The opponents united to found antivaccination leagues in Europe and the United States, and for decades they did strenuous battle with the public-health reformers who kept pushing for national vaccination programs. Eventually the public-health forces prevailed, but while those wars were under way, people by the millions were getting vaccinated privately. The information spread around the world, much more quickly than smallpox itself had. Most governments were slow to adopt universal vaccination programs, but they were also slow to adopt regulations restricting immunization, and people were free to vaccinate themselves, their families, and their patients if they had the general idea of how to do it and access to some cowpox. The spread of vaccination was on the whole beneficial, but of course there were plenty of mishaps, fatalities, and unsuccessful attempts along the way. In 1977 the last known case of smallpox—a man in Somalia—was reported.

So in a bit less than two centuries from the time Dr. Jenner performed the first vaccination, smallpox ceased to be one of humanity's chief scourges and became instead a strange medical curiosity. The last remaining stocks of the virus survived on microbiological death row, frozen in liquid nitrogen in two closely guarded laboratories—one in Atlanta, one in Moscow—while scientists and officials argued about whether it would be better to destroy them or to keep them alive for future research. The course of human evolution had been altered, with millions of people surviving who would not have done so otherwise, and so had the evolutionary career of the smallpox virus: it stood a good chance of becoming the first species that humanity had deliberately rendered extinct.[7]

During that 200-year period, many things changed. Immunology and epidemiology became important sciences with international networks of information exchange. The miasma forces savored a brief moment of what seemed to be final triumph in the 1820s when French doctors studied a yellow fever epidemic in Barcelona and concluded that the disease could not possibly have been spread by contagion—and then were forced into retreat when the convergence of microscopy and medicine enabled scientists to see germs with their own augmented eyes. Vaccination became widely accepted as a matter of public policy not only by national governments but also by international agencies such as the World Health Organization, and medical teams traveled about the globe immunizing people by the millions. The understanding of contagious disease changed fundamentally. People thought about it differently, dealt with it in different ways, were drawn into global webs of information and economics and public policy. And today, as the plague of AIDS spreads about the world, we expect and demand that somebody—the researchers, the doctors, the government—*do* something about it.

Immunology: The Quiet Explosion

When Jonas Salk developed a successful vaccine against polio in the 1950s, it was global news of the first magnitude. Today, we expect that new vaccines for previously unpreventable diseases will be announced periodically, and it is a steeper climb for any such announcement to make the front pages. One that did was the chicken pox vaccine, approved for use early in 1995, which caught the public's attention even though the disease only very rarely proceeds into serious complications such as pneumonia and brain inflammation.[8] An AIDS vaccine, if it ever comes along, will be the biggest scientific news item of all.

The situation as of this writing—situations change fast in the bio-information era—is roughly as follows: basic research—notably the

Human Genome Project—is proceeding rapidly, laying the foundation for continued progress in immunization against many diseases. This is true even though federal funding in the United States has been increasingly tight for years. If the political and economic climate were different, if the public purse strings were looser, basic research would be proceeding even more rapidly. Meanwhile, in the state-of-the-art laboratories of biotech companies in the developed nations, millions of dollars of entrepreneurial capital are being expended on research aimed toward developing new vaccines considered to have strong profit-making potential.

Elsewhere, teams of scientists—some well funded, some not—are working on vaccines against diseases that afflict people in less affluent regions of the world. The most heartening news on this front comes from Colombia, where Dr. Manuel Patarroyo appears to have developed a vaccine against malaria.

Malaria is one of the big killers, as serious as smallpox was in its heyday—and, like smallpox and polio, it is also a great destroyer of children. Sometimes called the "king of diseases," it strikes some 300 million people annually, killing a million children a year in Africa alone. There are many remedies for it but no sure cure, and no preventative. It is caused by a parasite that invades the bloodstream, colonizes the liver, and infects the blood cells—and parasitic diseases have been notoriously unresponsive to any kind of vaccine. For a while DDT eradicated malaria in some places—by killing the mosquitoes that carry the parasite—but then the mosquitoes evolved resistance to DDT, and the toll began to rise again. The impact goes far beyond the death count: many people suffer from malaria on and off throughout their lives, and the result is a chronically low level of health and productivity.

Malaria has been a political issue too, in its own way. Political and scientific leaders in developing countries have long suspected that a malaria vaccine was not regarded as a particularly attractive prospect by the private-sector research companies, and those suspicions were confirmed in 1984 when the World Health Organization sought the assistance of Genentech—a leading California biotechnology firm—in developing and marketing a malaria vaccine prototype. The company took a look at the prospects of making a profit for its shareholders on such an enterprise, and its vice president for research stated that "the development of a malaria vaccine would not be compatible with Genentech's business strategy." From a management point of view he was quite right, and for the next decade or so most of the serious research on malaria vaccine in the United States was done in projects financed by the Agency for International Development and private foundations. The Department of Defense also sponsored research, seeing the benefit of being able to vaccinate American troops stationed abroad.

Dr. Patarroyo's vaccine excited a great deal of controversy—partly because some of the test results indicated it was only about 30 to 35 percent effective, and partly because it has been hard for scientists in the United States and Europe to accept the possibility of a major scientific breakthrough emerging out of such an unlikely place as Colombia. International scientists and agencies such as the World Health Organization (WHO) argued bitterly about whether the vaccine even merited proceeding to large-scale testing. While the arguments raged, Dr. Patarroyo was hard at work on an improved model. A comment by an official of the Special Program for Research in Training in Tropical Diseases, an agency of WHO, says a lot about the rapid rate of change in the immunization field. Speaking in favor of proceeding with the first Patarroyo vaccine, he said: "It's a bit like buying a computer: You have to plunge in and take what is available now. The only difference is, you know for sure your computer will be obsolete a week after you've bought it. With a malaria vaccine, we've probably got more time to make use of it."[9]

Vaccine researchers look for ways to immunize people against previously unconquered diseases such as some kinds of cancer, look for better ways to deliver vaccines—such as nasal sprays or pills instead of injections—and explore new technologies such as "mimic viruses" constructed out of artificial protein fragments instead of the killed or inactivated actual viruses that in some cases have caused the disease they were supposed to prevent. Some are working on vaccines for other purposes: vaccines against male baldness and tooth decay are confidently predicted by a few futurists, and vaccination is a promising alternative approach to birth control. Chiron, one of the leading American biotechnology companies, has pioneered the development of "postinfection" vaccines that can reduce the severity of diseases such as genital herpes already contracted by the patient.[10]

You don't see vaccines at work, nor much evidence of them on the human body—save perhaps the mark of a smallpox vaccination—but, visible or not, the immunized body is a body that has been biotechnologically augmented, and is profoundly different from the one that is not. And growing numbers of people are going about their business with bodies that are augmented in other ways, structurally and organically.

The Bionic Body

Sometime in the distant past people began figuring out ways to compensate for injuries that would have been life-ending for earlier primates—a stick to aid the walk of a person who had broken a leg, a wooden stump to replace a lost foot. Ancient records tell of more ambitious substitutions such as artificial ears and noses. In more recent centuries people became accustomed to many devices that in one way or another attempted to compensate for the ravages of injury, illness, or age: eyeglasses, glass eyes, ear trumpets, false teeth, hairpieces, and increasingly sophisticated artificial limbs. In the 1970s television carried this to mythic levels with the Bionic Man, an astronaut who had been rather extensively deconstructed in a missile launch mishap and then miraculously rebuilt by scientists. "We can make him better than he was," they said in the voice-over introduction to each episode, "stronger . . . faster." They gave him a powerful mechanical arm suitable for throwing bad guys through walls, mechanical legs that could run faster than a speeding bullet, and a zoom eye that could lock in telescopically on distant objects. The only thing that seemed to be beyond the reach of science at that point was the ability to implant some acting talent into the star of the series.

The Six-Million Dollar Man and its equal-opportunity spinoff, *The Bionic Woman,* brought *bionic* into the popular vocabulary. We tend to misuse the word nowadays—technically it means combining biology with electronics, and most of the gadgets being implanted in human bodies are not electronic—but it really doesn't matter. The language we learned a few decades ago just didn't have a word to cover all the things that are now done to repair, improve, beautify—and in many cases save the lives of—human bodies. *Bionic* and the older word *prosthetic* (which, in many cases, isn't precisely correct either) will have to do.

Not long ago, during a bicycling weekend in the California wine country, I happened to get talking with an elderly man who mentioned that he had an artificial hip joint. He had ridden forty miles or so that day, and said he hadn't had any trouble. He mentioned that his other favorite pastime was square dancing. A real bionic man, sipping a glass of Chardonnay in a Geyserville hotel.

That millions of people now have artificial joints—shoulders, elbows, wrists, hips, knees, ankles—is not a particularly exciting piece of news nowadays. A bit more striking is the range of other mechanical devices now in use: Pacemakers—truly bionic instruments that regulate the heartbeat. Artificial sphincters, implanted near the bladders of men and women, which the wearer can operate by squeezing a small pump implanted in the scrotum or vaginal lips. Penile prostheses (not true prostheses because they do not substitute for the actual organ), which, either by means of a semirigid implant or implanted inflatable tubes, enable otherwise impotent men to achieve erections. Artificial inner-ear structures to replace delicate bones destroyed by chronic infection. Cochlear implants—in which electrodes are inserted into the cochlea of the inner ear—have restored hearing to thousands of totally deaf people. So far these are the most complex electronic devices implanted in humans, but miniature computers placed inside the eye are being tested now as a way to restore at least partial sight to the blind.

A new concept of medicine has evolved along with such gadgets. There used to be three treatment concepts: preventative (avoiding the illness), palliative (relieving symptoms), and curative (stopping the illness). Now there is a fourth: substitutive—finding ways to substitute either a body part, such as a joint, or a function, such as normal heartbeat.[11]

Not all of these devices can be made to disappear within the body, or to do their bionic business automatically and enable the wearer to live a normal life. Substitutive medicine has also produced huge creaking devices like the iron lungs in which some polio victims are confined for a lifetime, and the pioneering artificial heart that kept retired dentist Barney Clark alive for 112 days in 1983. The first artificial organ was the dialysis machine, an awkward but fairly effective device that Dr. Willem Kolff, the father of bionic medicine, began using on patients with damaged kidneys in 1943. Dr. Kolff kept improvising new models as he went—in the 1960s he made some good ones out of Maytag washing machines—and the technology has come a long way in its fifty-odd year history. Now something like half a million people are being kept alive on dialysis machines, and there are many models, including so-called artificial kidneys, that are actually portable dialyzers—but none of them is yet an adequate replacement for a living human kidney.

At this stage in the progress of substitutive medicine, cumbersome devices such as dialyzers and artificial hearts are most satisfactory when they are used as *temporary* substitutes—to keep patients alive until they recover from an acute illness, undergo surgery, or get an organ transplant. But everything is in motion, and this will change. Like many of the revolutions we are encountering in these pages, bionics is a very young revolution—yet it has become in its short life a respected branch of medicine, a growing body of scientific research, and also a thriving industry. There is big money in it, and new products are emerging regularly.

Some of the future products will be improvements on the old ones—better limbs and joints and pacemakers and implants, truly portable dialyzers—but more spectacular breakthroughs are possible. Some people predict the invention of artificial eyes and ears that will be not cosmetic substitutes, but actual working organs able to sense light and sound and transmit perceptions to the brain. Researchers are working on materials that will more closely resemble human bone and tissue, even be "resorbed" back into the body over time and replaced by natural bone and tissue. One such product is a paste that can be injected through the skin into a broken bone, where it hardens quickly into a substance that resembles natural bone but is stronger, and is gradually replaced by real bone as the break heals.[12] Replacement cartilage is a relatively "natural" augmentation, since it is created by growing the patient's own cartilage cells in culture and then injecting the tissue into a damaged joint.[13] Several laboratories are hard at work on possible blood substitutes, most of them made by coaxing the genes for human blood proteins into *E. coli* or other workhorse bacteria. Some skin substitutes are already on the medical market, and more are on the way. One such product is manufactured by taking cells from foreskins removed from baby boys in circumcision, combining the cells with purified collagen from cows, and culturing the resultant interspecies mixture in a nutrient medium.[14]

The future of bionics is partly predictable, partly unpredictable. We can predict that the science/industry/art of substituting human body parts and functions will progress—probably quite rapidly, although of course not rapidly enough for people who want the parts. We can't predict what materials, approaches, or technologies will work best. A potentially powerful dark horse in this race is organ regeneration—growing an entire organ from a few cells. One researcher says: "If, for ex ample, beginning with one's own cells, an individual's liver may be regenerated, the problems of organ rejection or those associated with the required immunosuppression would be avoided. Such a breakthrough would allow treatment of a number of diseases including pancreas replacement in diabetes and intestine regeneration in ulcerated colitis and related bowel disorders."[15]

My guess is that we will see more convergence—substitute body parts engineered partly out of living cells and tissue, partly out of mechanical and electronic elements. There are some signs of that happening already, but for now most of substitutive medicine is proceeding along one or another of two semiseparate paths: artificial devices and organ transplants.

New Hearts for Old

Bionics still has a faintly science-fictionish aura for most of us, even though the odds are good that somebody in your family has a pacemaker or an artificial joint. Organ transplantation—while an equally complex technological feat—lives in the realm of human drama. There is something about the passage of life from person to person, the organ taken from one body and put into another, that seems to get us where we live.

So we read, and are moved by, the story of the American parents who, after their seven-year-old son was killed by a stray bullet fired by bandits in Italy, arranged to have his organs donated to Italian children. In 1995 CBS ran a touching documentary entitled *A Heart for Olivia,* about the little girl from North Carolina who was the youngest person ever to receive a new heart—when she was less than two hours old. (Her potentially fatal heart malformation had been detected in the mother's fifth month of pregnancy; the donor was a baby born brain-dead in Spokane, Washington.)

The history of body transplants is shorter than that of bionics and prosthesis. There are of course legends of early attempts to transplant limbs, noses, ears, skin, and even internal organs, and some progress with transplant surgery was being made throughout the early twentieth century—but most efforts, even those that appeared at first to be successful, were stymied by the body's rejection response.

The recent amazing increase in the number of successful transplantations was made possible in part by advances in surgical techniques—many of them achieved with animals—and in very large part by advances in pharmacology. All too frequently a transplant operation would appear at first to be a resounding success—a transplanted kidney came alive almost magically, took on color, began to function—and would then turn into a disaster as the recipient's immune system detected and destroyed the foreign presence. New bio-information changed this. Medical scientists gained increasingly precise understanding of how the immune response did what it did. And they were blessed by the miraculous discovery of a drug that could be used to neutralize that response.

The drug, cyclosporine A, was developed by the Swiss pharmaceutical company Sandoz from an unusual mold, distantly related to the one that had produced penicillin a few decades earlier. Its discovery sounds a bit like a lucky accident—and in a way it was that—but it was also the result of procedures that had been designed to make such accidents happen. Sandoz scientists were under instructions to bring back soil samples from any place they happened to visit, and somebody returned from a trip to Norway with a bit of tundra. It was routinely analyzed, and found to contain a fungus that researchers thought might be useful as an antibiotic or antifungal agent—but the test results were not impressive. The fungus remained literally on the shelf along with countless other samples from here and there until one of the company's biologists discovered that it had the remarkable property of suppressing the immune reaction without the disastrous side effects that had been produced by other drugs. Even then it took a while to convince the management to start producing the drug and testing it with transplants. In all, thirteen years elapsed from the time the sample was brought back from Norway (1970) until cyclosporine was cleared by the Food and Drug Administration (in 1983). It soon became the company's third-best-selling drug worldwide—first in the United States, where most transplant surgery was being done. Cyclosporine, says medical journalist Mark Dowie, "may not yet have revolutionized medicine, but it certainly revolutionized transplantation, and transplantation may yet revolutionize medicine."[16]

Although cyclosporine is far from perfect, it was good enough to make transplantation a legitimate and increasingly common medical procedure, and to facilitate the development of all the other changes that had to take place for that to happen—improvements in surgery and in the various methods of stockpiling and transporting organs, new laws and regulations, changes in social values and public opinion, and new information systems to facilitate the connection between donor and recipient.

I imagine that, for some readers, my whole line of argument about the bio-information era creating new, global patterns of interconnection among people may seem rather abstract and remote. Unless you are a scientist, it is a little hard to get excited about internationally networked systems of gene banks and data bases. But when you consider the growing inventories of organs for transplants, the growing networks of information exchange, the growing number of people who are going through life with hearts, kidneys, corneas, blood, and tissue from other people, it is hard to avoid the realization that we are becoming interconnected in a way that no other species has ever been—not only with our machines, but also with one another.

Biofeedback

One of my more bizarre memories of the late 1960s involves the times when I used to wend my way down the Los Angeles freeways to a space research laboratory where I would then go into a dark, quiet room and sit for an hour or so meditating with electrodes connected to my head.

This was part of a research program being conducted by psychologists who were studying the ability of people to exert deliberate control over their brain wave patterns. As one of their volunteer subjects, my directions were to learn how to get into the "alpha wave" state—measurable on an electroencephalogram—which recent research had shown to be the state attained by advanced meditators. Unlike them, however, I had the advantage of an electronic feedback system, which produced an agreeable steady hum when my brain was giving out the right kind of vibes. I soon learned—as do most people who participate in such exercises—that I could, without ever quite knowing how I was doing it, get into the alpha state rather easily, and stay in it. I could control my brain waves.

I thought of those early adventures in biofeedback when I read recently about experiments in which people used their brain waves to move a cursor about on a computer screen, and even to control the movements of a flight simulator. The experiments with the computer were being done at the Wadsworth Center in Albany, New York. A subject with electrodes connected to his or her scalp would sit in front of the screen and look at a cursor and at a right-angle target located in one corner of the screen. The electrodes were connected to the computer, which was programmed to translate the faint electronic emissions from the brain into movements of the cursor. The subject's task was to will the cursor into the target—which the experimenter would occasionally move from one corner to another just to keep things interesting. The most adept of them were able to put the cursor into the target in 70 trials out of 100. In the beginning most subjects said they used visual imagery to move the cursor; they would think of something floating to nudge it upward, for example. After a while they didn't have to call up such images. They could just move the cursor, even while carrying on a conversation with someone else. But they didn't know how they did it.

At Wright-Patterson Air Force Base in Ohio, subjects in the Biocybernetics Project were wired up and placed in a kind of mock airplane, a large box with no windows. They learned to adjust their brainwaves (no hands on the controls) to send signals that would cause the simulator to roll from one side to another, as it would when making a turn. The idea was that with this technology, pilots may eventually be able to perform simple tasks such as changing radio channels while keeping their hands free for other things—or might even be able to fly a plane that way.[17]

Biofeedback—of which the above are but a few examples—entered the public consciousness in the 1960s, and tended to be identified with the *zeitgeist* of that era—the discovery of psychedelic drugs, the enthusiasm for Eastern mysticism, the feeling that the solution to just about everything was right around the corner, and the general giddiness about the untapped powers of the mind. But it also had good claims to scientific respectability. It used a machine, which was reassuring to American scientists who might otherwise have dismissed it as utterly flaky. And its effects could be measured and tested. It was not difficult to establish that patients could indeed learn to reduce their blood pressure, alter their skin temperature, or change their heart rate.

Biofeedback did not quite fulfill the immediate hopes of some of its more enthusiastic advocates such as Dr. Barbara Brown—who was conducting biofeedback research in Los Angeles at about the same time that I was one of its guinea pigs, and proclaimed it to be "the closest thing to a panacea ever discovered."[18] But "augmented proprioception," as it is sometimes called, has been shown to have many therapeutic uses for treating such problems as tension headaches, muscle spasms, speech disorders, phobias, insomnia, addictions, and impotence. One recent study concluded that "the scientific support for applied biofeedback has clearly increased, in spite of skepticism, criticism, and the complexities of conducting high-quality and meaningful research in this field."[19]

Biofeedback is a particularly fascinating example of Mazlish's fourth discontinuity, the breakdown of the boundary between human and machine. It would be easy to assume—as many do—that such connections are inevitably dehumanizing, lessening the power and autonomy of the individual. But with augmented proprioception, the power of the individual increases. Biofeedback has challenged earlier theoretical assumptions about how many bodily functions can be controlled by the conscious mind. It challenges those assumptions in practice for the stroke patient who is connected to an EMG (electromyography) device and learning to move paralyzed muscles. The machine does not move the muscle. It only provides bio-information to the person, who uses that information to regain control of his or her own body. This can be—if we understand it and make the most of it—part of a truly revolutionary change in human life.

The Knowledge Patient (and the Augmented Doctor)

Peter Drucker tells us that the information society has created a new kind of worker, a "knowledge worker," who simply cannot be managed according to the rules that applied in the day of the organization man and the blue-collar grunt.[20] Harlan Cleveland has written of the "knowledge executive" who needs to figure out the new rules—and figure out, too, that the new rules soon metamorphose into old ones.[21]

Similarly, we begin to find in the early years of the bio-information era that health maintenance will not operate as it did in the past. More areas of life are subject to human action, information, and choice. Not too long ago, diseases were things that simply happened to people, acts of blind fate or acts of God. There are still people who think that way—who see AIDS, for example, as divine punishment of sexual deviance or drug addiction—but most of us think of AIDS as a virus that may or may not be transmitted, depending on whether people practice safe sex and use clean needles. Goodness, as Mae West would say, has nothing to do with it; information has everything to do with it.

The bio-information revolutions are transforming medicine, and that transformation goes far deeper than you might gather from reading the news about the latest drug, vaccine, or therapy. Obviously, medical professionals are taking a lot of new information and learning new skills. Less obviously, they are learning in new ways: using computer models, for example, to peer into the human body and see how it works. They are being forced to become "knowledge healers"—not only learned in the old-fashioned sense but also adept at cruising through the networks and data bases when they need to learn something new, and aware of the rules about how information works and doesn't work.

Yet all of that may only be the tip of the iceberg, the visible but relatively unimportant part of the ongoing medical revolution. Much of what is going on now challenges the authority structures of medical practice, promises an end to the time when patients humbly offered up their bodies to be repaired, and doctors exercised godlike authority. There are, God knows, plenty of authoritarian doctors around, but they are dinosaurs. The best of contemporary medicine requires the active—and informed—involvement of the patient. Many of the products coming out of the biotech-infotech convergence aim in that direction.

The home pregnancy test, an early and relatively noncontroversial product of biotechnology, is one example of such involvement. Biosensors (electronic devices that can automatically measure the levels of certain substances) are another. Proponents of this technology see patients taking their own blood tests at home and transmitting the information to a laboratory, or people with chronic illnesses monitoring their own metabolisms.

Biomedical imaging and modeling are already being used not only to assist doctors and researchers and medical students in their work, but also to help patients understand the nature of their medical problems and to make better-informed choices about possible treatments.

An article in the computer-hip magazine *Wired* proposes HOT (health-oriented telecommunications) as a way to take this trend a giant step further. It notes that "of all the actions that make up this vast humble-jumble we call 'health care,' the great majority are already transfers of information rather than shots, cuts, and pills. The information may be diagnostic—such as X-rays, vital signs, and the patient's look and feel. It may be therapeutic—prescriptions and orders for tests. It may be consultative—'The pneumonia is one thing, doctor, but I don't like the look of these infiltrates.' It may be educational—'Mrs. Jones, for the good of your baby, you have to keep your weight up during pregnancy. This is not the time for a crash diet.' Or it may be about billing and insurance matters." The article also notes that, even in these high-tech times, much of this information "is handled in ways with which Charles Dickens would be completely comfortable."[22] The *Wired* solution is to set up new networks of information exchange for vastly expanded communication among patients, doctors, researchers, emergency workers, counselors of all kinds, networks of victims of similar disease or injury.

More networking. Clearly, the growth of bionic augmentation is a *social* event. It creates not only linkages between humans and machines, but all sorts of new interconnections among people. There is every reason to expect that—whatever specific forms the future of medical technology may take—this process will continue. And there is another important dimension to this evolutionary transition, one we have not yet examined. As human beings become different kinds of animals, their biological lives augmented in many ways, they also form new patterns of connection with other living things. The evolutionary shift now under way is brought about by human action, but it does not affect only human beings. It touches all life on Earth.

CHAPTER SIX

The Human–Animal Connection

Q: What kind of valve did they put into your heart? A: What they call a pig valve, because it comes from a pig. Every time I pass a plate of barbecue, I cry. It might be one of my relatives.

—Senator Jesse Helms[1]

All of life on Earth is interaction—from beginning to end, there are no truly solitary creatures—and the richness of life is expressed not only in the diversity of species, but also in the diversity of ways they cooperate and compete with one another.

There is, for example, the ancient drama of hunter and hunted—wolf and caribou, lion and wildebeest—in which the predator depends on the prey for its food supply, and in return performs the stern Darwinian function of weeding out the weak and maintaining limits on the size of the herd.

There are also neighborly interactions of mutual support. The cleaner wrasse, an admirable citizen of the Pacific coral reefs, sets up a sort of service station, and other fish line up there for the wrasse's attention. When the client fish's turn comes, it hangs in an unmoving, trancelike position with its mouth open while the wrasse goes over it—gleaning parasites from its scales and gills, even from the inside of its mouth. A wrasse will swim right into the mouth of a larger fish and work on its teeth without being eaten.

There are outrageous tricksters such as the species of blenny that opens up its shop in imitation of the cleaner wrasse. When a big fish comes by and goes into the passive pose appropriate for being cleaned, the blenny takes a bite out of its side and runs like hell. There are plant plagiarists such as the Ophrys orchid that has evolved a flower with patterns and growths to resemble a female bee, and even gives off a scent like that of a female insect, so as to lure a male bee into mating with it and accomplishing the pollination of the flower.

Then there are cases of true symbiosis, such as the arrangement between wood-eating termites and the protozoans that live in their digestive tracts. The termite has no genetic skill that enables it to get nourishment from wood, but the protozoans do: they produce an enzyme that digests cellulose. The protozoans can't survive outside the termite, and the termite can't perform its famous occupational specialty without the protozoans.

In any ecosystem all kinds of interactions such as these are going on at the same time, in complex and ever-changing patterns. The great majority of them are instinctive, developed out of the slow, mysterious trial-and-error process of genetic evolution. Plants and lower animals rely heavily on the genetic programs they were born with, while higher animals are more likely to relate to other species partly on instinct and partly on the basis of things they learned after they were born: a wolf pup learns that it is a good thing to bite a rabbit, not so good to bite a porcupine. Scientists are still learning about learning, and endlessly arguing about where the genes leave off and culture begins—but this much is clear: humans have developed a completely unprecedented range of relationships to other species, and most of these are *inventions*. And new inventions are coming along—indeed, appearing at a more rapid rate than ever before. The relationship between humanity and other animals is changing in fundamental ways, and this is a central feature of evolution now.

Us and Them

As human beings evolved in the past, they continually improvised new relations with other living things. As that happened, the course of evolution shifted for other species as well. The wild dogs around the fire became human companions, eventually were selectively bred to produce an astonishing range of creatures from Chihuahuas to Great Danes, from vicious pit bulls to the amiable but somewhat neurotic Australian shepherd that is at the moment under my desk scratching at a flea. The sheep and goats that were once hunted became domesticated, guarded, bred, and eventually herded away from their natural habitats to greener pastures. Ecosystems changed wherever the herds went, and predators became the enemies of pastoral peoples—which meant that their evolutionary fortunes were affected also. And so it goes. Waves of change ripple outward from each human transition, altering the world and everything in it.

Hunting and pastoral peoples depended on animals for far more than food. Skins became clothing and shelter, bones became tools,fat became fuel and medicine. And the relationships have always been deeply psychological as well as utilitarian: in rituals and myths, with totems and symbols, primitive peoples drew on the strength of animals, sought their protection, sometimes made them gods. Totemism is still with us in the form of the Detroit Tigers, the Golden Gophers of Minnesota, the San Jose Sharks, the comic bears and eagles that cavort on the sidelines during football games. Animals stalk, gallop, fly and slither through family names, coats of arms, military insignia, commercial brands.

But despite our intimate involvement with, and dependence on, all kinds of nonhuman creatures—or perhaps because of it—human beings have always harbored deep fears of becoming too deeply connected with animals, slipping into animalism themselves. Animals figure heavily in taboos as well as totems, are often devils as well as gods. In Christian folklore the Evil One is often shown to be half-animal, with tail and horn and cloven hooves. People in Medieval Europe were executed for the abominated (but apparently all too common) sin of sexual intercourse with animals, and often their poor partners in this crime were put to death as well.

The anxiety about maintaining the animal–human boundary was—much more than plain skepticism or religious piety—the source of the outraged opposition to Charles Darwin's theory of evolution. Although *On The Origin of Species* really said nothing explicit about a connection between apes and human beings, the general impression was that the book did assert such a connection—indeed, related humanity to all living things. "Our unsuspected cousinship with the mushrooms," was the way the prominent anti-Darwinist Bishop Samuel Wilberforce put it. (It was Bishop Wilberforce who, in a debate with Darwin's colleague T. H. Huxley, asked him whether it was through his grandfather or his grandmother that he claimed descent from a monkey. Huxley scored by replying that he would rather be descended from an ape than from Bishop Wilberforce.) That controversy, raging still, is essentially a border war, a battle to maintain a clear demarcation between human life and animal life. But although that demarcation may be clear in the minds of Bible scholars, it has never been clear at all at the level of our daily biological existence—and it is becoming, if anything, less clear all the time. It is one of the great ironies of the bio-information era that even as humans make an evolutionary leap that carries us further beyond any other animal, we become more intimately entangled with all the rest of earthly life.

Animal Farm

Although some uses of treated animal tissue in human bodies—such as Senator Helms's pig valve—are now quite routine, attempts to transplant whole organs from animals to human beings have so far not succeeded. Many people have tried, and sometimes made headlines for trying. One such case was the infant who became known to the public as Baby Fae—born with a fatal heart defect, then given the heart of a baboon. Baby Fae died 21 days after the operation, and the surgeon who had performed the heart transplant was severely criticized, not only by animal rights advocates, but also by many of his professional peers who felt that he had behaved irresponsibly—had, in the opinion of some, been so anxious to try his experiment that he failed to make an adequate search for a human heart.

But the availability of organs from human donors still lags far behind the demand for them; in the United States thousands of people die each year while waiting for suitable organs to become available. And some medical researchers are looking hard at the possibility of animal-to-human transplants on a large scale. "Now," reports *Bio-Science,* "biologists are rushing to develop a potential new source of donor organs: farm animals. Researchers envision organ farms, where pigs, sheep, and perhaps other animals may be raised not just for their meat but also for their major organs."[2]

The technical problem here—in addition to the obvious ethical, legal, and economic difficulties—is finding a way to get past the human immune system, which stands vigilantly on guard against foreign tissue. One of its lines of defense, the T cells, can be fairly effectively neutralized with immunosuppresive drugs such as cyclosporine when an organ from a human donor is transplanted. But when organ transplants from other species are attempted, the usual result is failure, hyperacute rejection, due to another weapon called the complement system—a fleet of proteins that move tirelessly through the bloodstream, clustering on anything they identify as foreign and marking it for destruction.

The complement proteins don't mark human tissue, because it has its own defenses—called shield proteins—whose specific job is to neutralize such attacks. All human beings normally produce these shield proteins, but other animals don't; that is the main reason why Baby Fae's body rejected the baboon heart.

The logical way around this problem is to produce genetically modified animals that have human shield proteins. A number of different teams are working along these lines, with considerable success.

One group in England has bred several generations of transgenic pigs. By the time this book is published, clinical trials with human volunteers will probably have begun. Many people who know they would soon die without an organ transplant would be willing to be part of experimental efforts that might save their lives, or at least allow them to contribute something to knowledge that could save others.

If this approach proves successful—and, sooner or later, it probably will—we then have a potential solution to the current shortage of organs for transplant. We also have a new chapter in the history of animal husbandry—and indeed in the history of life on Earth—because there has never been an animal able to exchange entire organs with those of other species. John Atkinson, one of the leaders in this field of research, sees pigs as particularly good candidates. "Pigs," he says, "are ideal in some ways, because their organs are the right size for humans, and the anatomy of the organs is very similar to man. Miniature pigs would be good for children with cystic fibrosis who need a lung, for example." He adds that pigs carry few diseases that can be transmitted to humans.[3] If this becomes common practice, then probably slaughtered pigs destined for market would first be "harvested," like the bodies of recently dead human donors, for usable organs. Or—as seems more likely, since viable organs are far more valuable than pork chops and bacon—pigs and other animals would be bred and raised for that purpose, and edible meat would be the by-product.

Animal Experimentation and Testing

Sometimes I discuss with friends the possibility that animal organs will be harvested for human use in the near future. A frequent reaction —even from people who wear leather shoes and don't mind an occasional steak for dinner—is one of repugnance at the idea that other living things could be so used, so exploited. Human transplantation, for all its ethical tangles, seems easier to take.

Yet over every human transplantation operation—over every sort of surgical procedure, for that matter—hover the ghosts of countless creatures large and small whose bodies have been sacrificed so that something might be learned. Every wonder of modern medicine is built on that living foundation. You do not receive a vaccination, get your teeth repaired, use a prescription drug, even powder your face or brush your teeth without drawing on a vast backlog of information obtained through animal experimentation and testing.

Nobody has a completely accurate count of the number of animals that have been used in the past or are used now; it runs into the multiple millions. A report from the Office of Technology Assessment estimated that 17 to 22 million animals are used annually in the United States.[4] Some representatives of the animal rights movement claim the number is closer to 100 million.[5] The Charles River Breeding Laboratories, also known as "the General Motors of animal breeding," produce some 14 million animals a year for the laboratories.[6] The uses are in such categories as the following:

- *Production of biologicals:* Animals of various kinds serve as living factories to produce biochemicals for use in research or therapy. I once visited a large commercial laboratory in Japan—the Hayashibara Company—where hamsters are used to produce large-scale multiplication of human cells for the manufacture of interferon.
- *Diagnosis:* Living animals have been the primary tools for use in diagnosing a wide range of human diseases including tuberculosis, diphtheria, brucellosis, and anthrax.[7] Until recently, rabbits were used to determine scientifically—that is, before it became otherwise obvious—that a woman was pregnant. The procedure was to take a sample of the woman's urine and inject it into the test animal; if the woman was pregnant, the sample would cause changes—observable upon dissection—in the reproductive organs of the rabbit.
- *Education:* Several generations of high-school students have gone through the ritual carving-up of frogs in biology class. A recent estimate was that over 3 million animals are used annually—frogs in high school, dogs for practice surgery in medical and veterinary schools.[8]
- *Toxicity testing:* One of the most controversial uses of animals has been testing the safety or limits of use of drugs, cosmetics, pesticides, food additives, industrial chemicals, and household products. A standard method has been the Draize test, in which concentrated solutions of the product are dripped into the animal's eyes, sometimes over a period of several days. Rabbits have been most commonly used for this test, which is gradually being phased out, but still encouraged by the U.S. Food and Drug Administration and the comparable agencies in many other countries.[9]
- *Research:* Animals of all kinds and sizes, from fruit flies to chimpanzees and of course including many mice, cats, and dogs, are routinely used in a wide range of basic and applied research.

For a variety of reasons the whole animal experimentation picture is changing—but living animals are still the instruments of choice in many scientific and technological procedures that help to make our lives more bearable. And we make our lives more bearable still by not thinking or knowing too much about the details.

One day some years ago I sat in at a meeting of young journalists, aspiring contributors to the news service I often write for, and two big issues came up: AIDS and animal rights. Some gay activists were present, and there was much talk about the book that had recently been published by Randy Shilts—*And The Band Played On*—condemning the medical establishment for its failure to move more effectively against the AIDS epidemic. Everybody agreed that the scientists and the government were at fault, and that more should be done to combat AIDS. Then later the talk turned to the animal rights movement, and everybody expressed strong support for it as well. It wasn't until afterward that I realized this feast of agreement really should have turned into an argument, because AIDS research is a prodigious consumer of laboratory animals—including intelligent higher primates, our closest relatives—and an accelerated research effort would inevitably lead to a higher cost in animal lives.

Of Mice and Men—and Man-Made Mice

In 1980 the word *transgenic* did not exist, even though *E. coli* bacteria with human genes were already hard at work producing human insulin. But a few years later scientists began creating laboratory mice with genes for human illnesses, and soon the word was a common part of the scientific vocabulary. The U.S. Patent and Trademark Office granted Harvard University the first patent on a transgenic animal—the so-called "oncomouse," which carried genes that made it susceptible to cancer—in 1988. More patent applications followed, and research in this area grew rapidly. In 1990 alone more than 600 scientific papers were published describing man-made mice with symptoms of human diseases ranging from arthritis and diabetes to obesity.[10]The procedure—which is rapidly becoming more precise—is to remove fertilized eggs from a mouse, get the human genetic material into the embryos, and then implant them in a foster mother.

By this method, researchers have created mice that carry the gene for beta-amyloid protein, the main component of the plaques that develop in the brains of people with Alzheimer's Disease. The patent rights on this particular mouse are held by Athena Neurosciences, a San Francisco firm working in partnership with Eli Lilly & Company, the giant international pharmaceuticals company. There is no certainty, of course, that having a mouse with Alzheimer's-like brain plaques will lead to a cure for Alzheimer's. There is no certainty, in fact, that the plaque is the cause of Alzheimer's; it might be the result. But the mouse offers the opportunity to test hypotheses and possibly develop treatments, and the stakes are high; analysts say an effective Alzheimer's treatment could easily be worth $1 billion a year.[11]

Animal Rights and Animal Wrongs

Contemporary animal rights organizations have historical roots in nineteenth-century England, where antivivisectionism emerged as a major social force with strong links to the antivaccination movement. These causes were part of the political and psychological turmoil of a society grappling with—among other things—the Darwinian assault on the biblical worldview and the advent of modern medicine.[12] The oldest such organization was the Royal Society for the Prevention of Cruelty to animals, founded in 1836 (Darwin was a member); its non-Royal American counterpart was founded in 1866. Eventually, many other such associations came along. And today, in the age of the organizational population explosion—an unexpected and politically turbulent side-effect of the information era—animal rights groups abound. The list includes the Animal Rights Network, the International Primate Protection League, the British Union for the Abolition of Vivisection, People for the Ethical Treatment of Animals, Trans-Species Unlimited, and the Animal Legal Defense Fund. There are large international groups with publications, direct-mail fund-raising campaigns, and Washington lobbyists, and there are small guerrilla groups in many communities and on many college campuses—some of which have little patience with the large ones. Their methods range from relatively sedate public education programs to raids on laboratories.

The animal rights people have driven a lot of scientists and university administrators crazy, and they have made a difference. Several kinds of difference, in fact. One writer on the subject reported:

> Recently—largely owing to pressures exerted by animal welfare lobbyists—Revlon and Avon, two giants of the cosmetics industry, have awarded very substantial grants to scientists at Rockefeller University for the express purpose of stimulating research into alternatives to the Draize test. Almost at the same time, the Cosmetics, Toiletry and Fragrance Association awarded a grant to the Johns Hopkins Center for Alternatives to Animal Testing; Bristol-Myers also funded alternatives

> research by the British group Fund for the Replacement of Animals in Medical Experimentation (FRAME); the American Fund for Alternatives to Animal Research (with the support of other animal welfare groups and individuals) initiated a similar project at the Medical College of Pennsylvania; and the New England Antivivisection Society did likewise at Tufts University School of Medicine.[13]

Largely in response to the public pressure generated by animal rights activists, laboratories everywhere have become much more scrupulous in their treatment of animals—and also of animal-rights activists, whose opinions are frequently sought. This presents something of an ethical dilemma for the animal-rights people, who generally oppose all experimental use of animals yet find themselves tacitly supporting it by involving themselves in the process. Some groups refuse to take part in any such activities; others, such as veterinarian Michael W. Fox of the Humane Society of the United States, believe it is necessary to do so.

> While there are many areas of laboratory animal exploitation and suffering that need to be prohibited, such as military weapons testing and extensive burn studies, as long as the public consensus accepts the use of animals for essential biomedical purposes, we owe it to the animals in the name of both compassion and good science to ensure that they are treated humanely (analgesics and tranquilizers being used whenever necessary) and kept in conditions that guarantee their overall physical and psychological well being.[14]

I served for a couple of years as an outside member of a review committee at Lawrence Berkeley Laboratories, where—along with a team of veterinarians and laboratory administrators—I went laboriously through the protocols that researchers were required to submit: numbers and types of animals to be used, exact nature of the experiment, expected outcomes and justification, method of anesthesia. Those descriptions of routine use of mice and other creatures would not make any animal lover happy, but they signified some real progress from the days when surgeons practiced on unanesthetized animals.

That's one way science can respond to the animal rights movement. Another way is to fight back, and it is an option that some people and groups in the scientific world find most attractive. There is now a fairly strong counter-movement. You can find its arguments expressed in polemical books such as *Man and Mouse: Animals in Medical Research,* by Oxford University pharmacology professor William Paton, who calls the activists "animal hooligans," and considers their philosophical arguments to be uninformed and muddleheaded.[15] You can, in fact, find organizations whose entire purpose is to counter animal-rights activism. One of the basic rules of life in the information age is that if you are being crowded by an organization, the best thing to do is form your own. I regularly receive in the mail a newsletter from one such, the California Biomedical Research Association—a coalition of scientists, university administrators, private companies engaged in laboratory research, and groups dedicated to the elimination of certain diseases. Its pages are filled with refutations of irrational or inaccurate statements recently made by animal rights activists, reports of attacks on laboratories, and news about progress being made—with the use of animals—against diseases such as AIDS and cystic fibrosis.[16] The animal rights movement—and the counter-movement—are in many ways exemplary of information-age politics. We have the proliferation of organizations, the erosion of secrecy—experimenters can no longer assume that what they do will be protected against public exposure—and a stunning variety of viewpoints. On one side are the ethicists and activists who believe that—as Peter Singer, one of the movement's leaders, puts it—"nonhuman animals are an oppressed group" that we treat "as if they were things to be used as we please, rather than as beings with lives of their own to live."[17] For them, the only solution is a revolutionary shift of consciousness, and a complete abandonment of any activities that involve the suffering or discomfort of animals. On the other side are the many scientists, political theorists, victims (and relatives of victims) of disease, and others who think that the animal-rights cause is fundamentally flawed and that its leaders are a bunch of posturing troublemakers. And somewhere in the middle are the reformers who work for such goals as alternatives to the use of animals for research, better ethics education for scientists, and more review committees on animal experimentation.

This is indeed another information-age phenomenon, a convergence—this time more like a collision—of advancing scientific and technological skills that use (and create new uses for) animals; and a growing body of information, sometimes all too vivid, about the pain and suffering such uses involve. It is also a part of *Homo sapiens'* present evolutionary transition. No other species has developed such a range of uses for other living things, and no other species has developed an institutionalized capacity for feeling guilty about it. I suspect that as long as the uses and practices are a part of human life, the guilt will be also.

Paths Beyond the Present

Does bio-information progress presuppose ever-greater exploitation of animal life, with never-ending political, ethical, and religious controversy about the pain and suffering that we impose upon other living creatures by our ingenuity? Or does it offer hope of liberation from such uses, point toward a future in which human beings will gain access to the whole vast library of biochemical skills and strategies encoded in animal genomes without doing harm to the animals themselves?

We have just considered some of the exploitations, and will examine several more in later chapters. Consider, on the other hand, some of the early products of biotechnology that lead away from animal exploitation.

The first genetically engineered item on the market was Humulin, short for human insulin. The history of successful treatment of diabetes—or even knowledge about its cause—is a good deal shorter than you might think. It was late in the last century—1889 to be exact—when a pair of German scientists demonstrated, through experiments with dogs, that removal of the pancreas caused all the symptoms of diabetes. Just after the turn of the century, an American scientist demonstrated the connection between diabetes and specific cells in the pancreas, known as the islands of Langerhans, and a British scientist proposed that these cells secreted a substance—not yet named—that controlled the metabolism of sugar and other carbohydrates in the body. In 1921 a Canadian doctor isolated this substance—through more experiments on dogs—and named it *insulin* after the islands. That was the beginning of insulin therapy—treating human diabetics with insulin taken from the pancreases of slaughtered cattle and pigs.[18]After that, two interesting and apparently contradictory things happened. One was that insulin therapy became a widespread medical practice, enabling countless diabetics to live relatively normal lives. The other was that the number of diabetics increased, and many people still died of it. Some antivivisectionists have suggested that this demonstrates the failure of insulin therapy.[19] However, there's persuasive evidence that the main reason for the increase was that more people lived longer, as a result of other medical advances, and survived into the years when they became subject to late-onset diabetes. (One of the ironic inevitabilities of medical progress, as many people have pointed out, is that it enables you to live long enough to die of something else.) Another possibility is that insulin therapy, by enabling young diabetics to live longer and have more children, increased the number of potential diabetics in the human gene pool. This is what is sometimes called reverse eugenics, or dysgenics.

In any case, the market demand for animal insulin has increased steadily over past decades. However, we now have an alternative, Humulin, which is not taken from slaughtered animals. Although its manufacture does involve the use of living things—i.e., bacteria—this is a form of exploitation to which most animal-rights activists do not object. Biotechnology has cut one chain of human-animal dependency. It has done so, also, in the case of Chy-Max, the artificial rennin now used in the manufacture of cheese instead of the substance taken from calves' stomachs.

There are two reasons why animals are now used less often for medically related purposes. One reason is that animal-rights activism has had an impact on public opinion, causing some schools to give up on frog dissection and some companies to proclaim that no animals are sacrificed in testing their products. Another reason is that many of the uses are becoming obsolete. It is no longer necessary to use rabbits for pregnancy tests, and kits based on monoclonal antibodies are now able to produce much faster and more accurate diagnoses of many diseases than the older methods involving animal use.

Cell and tissue culture provides an effective alternative for testing toxicity. A sample of human cells does not always provide enough information to tell how an entire organ or a living person might respond to a certain substance, but it does get around the problem of species variation—the possibility that the test animal might have a greater or lesser resistance than a human being would.

Another potential alternative—more dramatic in its promise, less advanced in reality—is computer modeling: artificial-life programs that so precisely mimic the complexity of living things (or "wet life," as the modeling enthusiasts call it) that they can be used to test possible therapeutic substances and strategies.[20]

"Alternatives" is a favorite buzzword in the animal-rights dialogue, and there is real progress in that direction. But no end to the use of animals in medical research is in sight, and in fact some of the alternatives are being developed by using animals. As the great British biologist, Sir Percy Medawar, once observed: "We must grapple with the paradox that nothing but research on animals will provide us with the knowledge that will make it possible for us, one day, to dispense with the use of them altogether."[21]

CHAPTER SEVEN

The Body Politic: Private Lives and Public Issues

We are going to have to redefine our notions of what motherhood, fatherhood and pregnancy are. Some women will become biological mothers but won't get pregnant, hiring a birth mother instead. Other women will choose to become pregnant later in life, after their careers and even after menopause, either by carrying their own embryo conceived years earlier or by buying eggs from another woman. Some women will become pregnant without meeting the father. Men will become biological fathers without meeting the mother. Some babies will be born without anyone becoming pregnant at all! And there are other permutations as well.

Undoubtedly, many of us will initially find these ideas too abhorrent, too foreign to the basic human way of life. What right do science and medicine have to intrude upon this most natural of life processes and make it so unnatural? This sort of ethical question is going to come up again and again in the future of health care, and not just in the area of reproduction.

—Dr. Jeffrey A. Fisher[1]

The scientific and technological revolutions we read about in the newspapers soon come into our homes and begin to alter the ways we manage our bodies and live our private lives. People everywhere are taking in new information about nutrition and health maintenance, learning about organ transplants, becoming acquainted with genetic screening and artificial insemination, deciding about abortion and euthanasia: making choices that are often literally matters of life and death. We have a growing number of such choices to make, and we don't seem to have any choice about whether or not we will have more choices.

The choices increase personal power, and it would be easy to assume that they also increase personal independence—but that isn't exactly what is happening. As we have seen, entirely new kinds of *inter*dependency emerge. The human species is being drawn into new webs of connection to all of nonhuman life, and each individual person is being drawn into new webs of connection to other people. Augmentations are social events. The biological systems we call our bodies hook up to other systems and subsystems: information networks, markets, organizations, governments. Every augmentation of the human body costs money—often a lot of it—and every biological choice we make seems to have either political implications or ethical ones. Evolution is turning out to be a great generator of controversy.

So much of private life becomes entangled with public policy. Those life and death choices are affected by court decisions about abortion, by the availability of birth control materials and sex education, by medical insurance, by regulations about the practice of medicine, by laws about whether people can decide when to die. Governments regulate sperm banks and abortion clinics, grant (or withhold) approval on new drugs and medical technologies—too quickly for some, too slowly for others. And even when we stay clear of the legal and bureaucratic machinery, we are still caught up in other webs, networks, and systems; we have to deal with the values and role expectations of one or more cultural heritages, we feel the pressure of peers and parents, we connect to markets and information centers.

Who Owns the Body?

Organ transplant is a particularly sensitive subject, both personally and politically, for many reasons: its relative newness as a field of medicine, the intense demand for organs and tissue, the potential for high profits, the need for efficient and effective methods, the danger to the patient from a diseased or damaged transplant—and, by no means least, the powerful emotions it engenders among families of donors, families of recipients, families of people who for one reason or another fail to find a transplant.

A person regarded as a desirable organ donor—a young man declared brain-dead after a motorcycle accident, for example—enters into a legal and biological status that until fairly recently simply didn't exist. His body is now called a "neomort" or "biomort"—inelegant terms that modern medicine has coined to describe a human being no longer alive but not yet fully dead. It is maintained in this state while some member of the hospital staff performs the distasteful task of approaching the family about the possibility of allowing him to become a donor. (That at least is the case in the United States, where the body is regarded as private property. In some countries the laws permit a hospital or transplant surgeon to take organs from the body of any person who hasn't written a will or declaration to the contrary. In others the law simply declares that at the moment of death a person's organs become the property of the state.)[2] If permission is given, different surgical teams begin "harvesting" the body—removing heart, kidneys, liver, corneas, mandibles, inner ear, and in many cases some portions of skin, bones, muscle tissue, cartilage, pericardium and the heavy brain-cover membrane called dura mater.

The organ-donating neomort enters into a set of human relationships that the living person had never imagined. The most important of these connections, of course, are with the prospective recipients. In the United States, the information on the available organs is immediately typed into computers at the hospital and relayed to the central data bank of the United Network for Organ Sharing in Richmond, Virginia. This data bank is in turned linked to the eight hundred or so transplant centers around the country, and to their patients—that word seems particularly charged with meaning in this context—who are awaiting transplant surgery. Various tests—including in some places DNA analysis—are used to determine the compatibility of donor and recipient. Some organs are immediately packed into coolers, rushed by ambulance to an airport, and then taken by another ambulance to a hospital where a surgical team is already at work preparing a patient. Other parts of the body go to private institutions whose business it is to preserve them and distribute them to hospitals. The private institutions are inspected and certified by private organizations such as the American Association of Tissue Banks and the Eye Bank Association of America. Various federal regulatory agencies are also involved. Organ and liver transplants are managed by the Health Care Financing Administration and Health Resources and Services Administration; bone marrow transplants by the National Institutes of Health; dura mater, heart valves, and corneas by the U.S. Food and Drug Administration (FDA) under the Medical Device Amendments of 1976.

One of the great fears is the possibility that some disease—such as AIDS—might be carried to the organ recipient. Testing procedures have improved, and the FDA reports a high rate of safety in the estimated 300,000 transplant operations now performed annually in the United States. But you hear horror studies such as the one concerning a man who died of gunshot wounds in 1985, and from whom 61 organs and tissue grafts were taken. The donor's blood had been analyzed twice for infectious diseases before any organs were removed, and in both cases the tests had been negative. Then, a few years later, using more refined methods, the tissue bank discovered that the donor's blood tested positive for HIV. They were able to locate thirty-four of the more than forty patients who had received the man's organs and tissue. Seven of them had contracted the AIDS virus, and three had died.[3]

This is a distressing subject, but it is more or less manageable both scientifically and politically. Scientifically, the procedures for guarding against such events are getting better. Politically, there are regulations in place. It is even, in a sense, manageable culturally. Most people have come to accept organ transplant, folded that strange reality into their views of what is practical and acceptable in the world, accepted its risks. But the course of events keeps testing the limits of social acceptability. Each time I read an article about neomort harvesting, the list of usable parts seems to have lengthened. And it may lengthen further yet: Mark Dowie, a sober commentator on this subject, thinks hands and feet will soon be successfully transplanted. Some of the surgeons who perform sex change operations are talking about sex organs: "I don't think the government will fund penis transplants, but we'll try to persuade it to," says one of the pioneers in this field.[4]

Another controversy concerns anencephalic babies—born alive with a brain stem but no brain, destined never to become conscious and to live no longer than a few days. About one thousand such babies are born each year in the United States, and in recent years the courts have said they are covered by the Americans with Disabilities Act, entitling them to the best medical care available. But they are also prime candidates to become organ donors, giving their healthy but doomed hearts, livers, and kidneys to other children. The American Medical Association has taken the stand that anencephalic babies should be harvested for transplantation while still alive. One ethicist called this a "horrific and horrendous idea."[5]

One is tempted to ask the old question—when will it ever end?—and no answer is immediately apparent. The question we must grapple with, day by day, is how we handle such matters—what values, what views of reality, structure our decisions. The idea of human parts circulating about through society is disturbing, but as I contemplate this emerging reality I look back—not forward—to the ancient practice in Tibet, where the bodies of dead people are not sealed off in perfumed vaults, but are generously handed to the world. They are taken to open areas where they are hacked apart, pieces offered to the vultures—the idea being that we came from the world, and go back to it. One of the newest connections has something in common with one of the oldest.

Choosing Your Sex and Gender

Nothing illustrates the present proliferation of personal options, and the social complications that come with them, more dramatically than sex change—once restricted to the realm of fantasy and now sufficiently common that a recent newspaper article includes "transsexual" among the words we must all add to our vocabularies.

Actually, the word was coined in the early 1950s by a psychiatrist describing the case of a girl who was obsessed by the desire to become a boy; he called the condition *psychopathia transsexualis.*[6] But at that same time, advances in surgery—building on reconstructive skills developed during World War II—and the use of hormones were making it possible for a person to realize those fantasies by being medically transformed into a reasonably good replica of the opposite sex. In 1952 the whole world heard about the case of a man named George Jorgensen who, with the help of Danish physicians, had been trans-sexed into Christine Jorgensen. The Jorgensen case was the occasion for much merrymaking in the media, but it was taken very seriously by men and women who felt they had been born into the wrong body—and it was taken seriously by professionals in psychotherapy and medicine who believed that surgery, in combination with hormone treatments and counseling, was the solution of choice for the condition that many of them preferred to call "gender dysphoria."[7]

Over the next few decades this evolved into a recognized branch of medicine. In 1967 the Johns Hopkins Gender Identity Clinic was opened in Baltimore, and dozens of other major university hospitals soon followed suit. Less than ten years later *Newsweek* reported that over 3,000 transsexuals in the United States had undergone surgery; it also mentioned that 10,000 more people were potential candidates, since they considered themselves to be members of the opposite sex.[8] Because the hospitals in the United States require an extensive period of evaluation and counseling for prospective transsexual surgery candidates, and generally end up performing surgery on fewer than 10 percent, many seek surgery abroad.

Most transsexual operations are male-to-female, but progress is being made in the opposite direction. In many cases female-to-male transsexuals merely get hormone treatments that gradually change their body structures and stimulate the growth of facial hair. Some proceed to what is known in the transsexual subculture as "top surgery"—mastectomies—and some go all the way to "bottom surgery," which involves various methods of constructing an artificial penis. The operations never produce a fully functioning male or female. The post-operative transsexual has to face a multitude of social and legal problems, ranging from "toilet trauma"—i.e., figuring out what to do in a public bathroom—to the difficulty of getting a new driver's license to the threat of imprisonment for violating the cross-dressing statutes that exist and are enforced in some places. Only rarely will insurance cover the costs. As one report puts it: "If your insurance company is persuaded that you truly have the psychiatric disorder of transsexualism, for which surgery is a necessary part of the treatment, you might get reimbursement from them—after you've agreed to go through life with an official diagnosis probably comparable in many people's minds to necrophilia."[9] But the transsexual urge is sufficiently strong that thousands of people have undergone great stress, pain, and expense on such treatments, and undoubtedly thousands more would gladly make the leap if the treatments were less expensive and more easily available. Transsexuality illustrates some of the realities of life in our time. One has to do with information: if an option such as sex change exists, people will find out about it. Another has to do with economics: if the demand exists, somebody somewhere will supply it —to those who can find the money.

It is very easy to keep a judgmental distance from transsexuality —to be amused by its weirdness, repelled by the carnal details of reconstructive surgery, cynical about the doctors who make a handsome living doing it. And I have to confess that I have felt a bit of all of those in the course of investigating this subject. But if you read a few accounts of people who have made these choices—read their passionate declarations of need to change sex, the stories of how they bravely deal with the consequences—you can't help feeling a touch of admiration for their courage, and also for the dedication and plain ingenuity of the people who have created this new branch of medicine. They are all out there on the evolutionary front lines, and in some ways their awareness of what is happening is more highly developed than it is for the rest of us. You can't experience sex change and not know that, in some fundamental ways, the world is changing also.

Even the startling reality of a man becoming a woman—or vice versa—may soon begin to seem a bit dated. It is, after all, based on the assumption that there are only two sexes, and very few assumptions about biological destiny are entirely safe these days. A recent book by a feminist historian presented a persuasive body of physiological data to support her argument that there are actually five distinct biological sexes among human beings (the two standard ones and three forms of hermaphrodism) and that sooner or later our culture is going to have to get past its present hang-ups on that subject. When that happens some people will be male, some people will be female, some will have sex-change operations in one direction or the other, and others will choose to come out of the closet and define entirely new social and sexual roles.[10]

Not long after I heard about that idea I appeared on a panel with a group of futurists, one of whom said the concept of five sexes was basically sound but the number was a bit conservative. She went on to suggest that if we paid more attention to other factors beyond mere physical differences—expanded our vision to take in all the known psychological types, social life styles, and genetic variations that influence sexual behavior without producing different sex organs—we would find people sorting themselves into about forty-seven different sexes and/or genders. She whimsically predicted that buildings would have that many restrooms—not what I would call a high-probability scenario, but we do face a very real prospect of fundamental changes in how people think about sex and gender. This is partly just a matter of new information circulating around through society, but it is also helped along by technological advances. Recent developments such as sex-change operations and the use of hormones further complicate the picture, and add "choice of sex" to the list of life-style decisions available to many people.

We find then that as the human species evolves, it is developing a rather startling range of sexual and gender possibilities and choices —considerably more than exist among other primates—and that these set in motion far-reaching social and cultural changes that, sooner or later, touch us all.

Parental Sex Choice

Meanwhile, many other people—ordinary people who would never give a thought to changing their own sex—are making choices about the sex of their offspring. People have been doing this for centuries by means of infanticide, simply disposing of a child of unwanted sex—usually female. But now technologies such as amniocentesis and ultrasound imaging make it possible to predict the sex of an embryo or fetus early enough for an abortion, and even more reliable—and less stressful —methods of sex choice will be developed in the near future. It is technically possible now for prospective parents to make a certain and reliable sex selection if they use *in vitro* fertilization. These choices, however made in the past, present, or future, are never really private choices either. They are ringed about by laws and regulations, public policies, cultural values, and social conditions that make certain options readily available and others unavailable.

In China they are powerfully affected by the convergence of the one-child-per-family policy with the traditional cultural bias toward male children. A report from Stanford's Morrison Institute for Population and Resource Studies forecasts a huge future "marriage gap" by the year 2020—a million men a year reaching marriage age without being able to find a bride, the inevitable result of the present ratio of 114 boy babies for every 100 reported births of girl babies. The report also points out that over time, this could have an ironic boomerang effect on genetic evolution in China. Couples who produced girls would marry off their daughters and have their lineages carried on, whereas families who produced boys would die out. Eventually—if the present practice continued over 200 generations or so—the Chinese population would develop a genetic bias toward girls, and more of them would be born.[11]

That isn't likely to happen, of course. Nor is it likely that Chinese males will solve the problem by going in heavily for sex-change operations. The likely result will be yet another increase in international migration, as single men go elsewhere in search of wives and single women go to China in search of husbands. And what should happen over time will be changes in values: higher social status for women, parents less obsessed with the goal of the cherished boy-child. But do not expect people to stop choosing the sex of their offspring. That isn't the way things are moving.

Birth Control Now

Dr. Jeffrey Fisher, one of the few medical people willing to make outright predictions about what is going to happen and when, says that reliable home testing of ovulation will soon make effective birth control available to people who are opposed to contraception.[12] This is, I think, a particularly intriguing peek into the not-too-distant future, because it is one of those times when the boundary blurs between high technology and plain information. If you want to, you can take a detour around the fancy birth-control technologies and use a method—itself the product of advanced biotechnology—that enables you to do family planning "naturally."

The ground rules of human reproduction have changed dramatically over the past few decades, and they are changing more rapidly now than they ever have before. There are more and more ways to avoid having a child, to have a child, or to select what kind of a child you will have.

In the field of birth control we have a wide range of options including long-term contraceptives such as Norplant, a capsule placed just beneath the skin of a woman's upper arm and effective for up to five years; intrauterine devices; barrier methods such as the familiar condoms and diaphragms and the newer vaginal condoms; "contragestives" such as the controversial RU-486, which can be taken after conception to prevent the fertilized egg from implanting in the uterus; and abortifacients such as the drug prostaglandin, which has the effect of inducing premature labor. In 1993 the Upjohn pharmaceutical company launched an advertising campaign for Depo-Provera, an injectable contraceptive that has been in use for some time in developing nations, and was then being offered to up-scale American consumers as "birth control you think about just four times a year."[13]

The list of alternatives keeps growing. Some time ago I read about an entirely different approach, patented by a New York gynecologist, that kills sperm with low-level electricity. A tiny battery, which can be left in place in the cervix for up to a year, produces less electricity than a pacemaker but enough to prevent conception. Its inventor said it had proved 100 percent effective in tests with animals; it had not yet been approved for human use and I don't know if it ever will be. But other inventions—including a pill for men and long-term but reversible sterilization—can be expected to gain regulatory approval and market acceptance. The main obstacle may turn out to be the entrepreneurial instincts of lawyers—some of whom have made a specialty of suits against Norplant by users claiming undesirable side-effects. The picture is complex; yet, for all the chaos and confusion of our biologically turbulent times, one thing is certain: the variety, availability, and effectiveness of birth-control devices and methods will continue to increase.

All these devices—the ones we already know about and use, the ones on the way—have that paradoxical character of simultaneously increasing the power of the individual and drawing us into new social/political tangles.

The ability to manage one's own reproductivity—which derives from information about how reproduction works—is an amazing augmentation of the power of the individual human being. Yet, just as vaccines turned diseases like smallpox into social issues, birth control methods turn pregnancy into a social issue. Abortion is the subject of one of our angriest political arguments. We also argue about sex education, about availability of birth control information and materials, about approval of RU-486. More subtly, we are challenged to revise our values about what is an optimum family size. I grew up in a world in which the large family was regarded as the epitome of the good life, and—somewhere in the 1960s—ran into an emerging value system, influenced by environmental concerns and the changing roles of women, that preferred small families. So, coming from a family of six, with one grandparent who came from a family of eleven, I am the proud and politically impeccable parent of an only child. Times change.

Reproduction Now

For aspiring parents who have not managed to conceive or complete a pregnancy, the list of options is growing and the cultural/political climate changing.

A lot more than preference in family size has changed in the past few decades. Less than forty years ago, a court in Illinois ruled that artificial insemination by donor—with or without consent of the husband—constituted adultery on the part of the mother, and that any child born as a result would be illegitimate. Today artificial insemination by donor is performed routinely, and there are well over a quarter of a million Americans who came into the world this way. We have grown accustomed to *in vitro* fertilization, to your friendly neighborhood sperm bank. Social norms and institutions slowly make room for single parents, gay parents. Surrogate motherhood is still controversial, but I expect it will find its place in the expanding repertoire of family life. Egg-freezing is likely to be perfected to a point that will make egg banks as commonplace as sperm banks are now. So, at the same time that we get used to a steady growth in birth control methods, we get used to more ways to have babies. The overarching trend is always toward more choices. Human reproductivity has become fundamentally different from reproductivity in other animals, and the change is taking place in what is, from an evolutionary perspective, little more than the blink of an eye. Our values and practices are different from those of our grandparents, and the values and practices of our grandchildren will be quite different from our own. That much we know—but we don't know how different, or in what ways.

The changes in birth control methods and reproductive methods are both striking and—to many—disturbing. But neither of those is quite as sensitive as deciding whether or not to have a given child, or deciding what sex it will be. It is becoming easier to make such choices, and more people are making them. Prospective mothers, especially older ones, routinely use amniocentesis or another prenatal testing method to determine whether their child may be born with Down's syndrome or some other genetic defect. Parents now are likely to know the sex of their child before it is born. *In vitro* fertilization makes prenatal testing much easier and highly specific.

Many people believe we should not have such choices, and that some (or all) of the birth-control and reproductive techniques should be outlawed. But outlawing anything of that sort is also a choice—a political one, not easily made and not easily enforced. Recent events in Italy illustrate this. In the early 1990s Italy became known for its rather freewheeling legal climate concerning reproductive matters. Although home to the Vatican, it was also the country with the lowest birthrate in Europe—strongly suggesting that a lot of people were practicing birth control—and it had a thriving business in all manner of reproductive alternatives such as artificial insemination and surrogate motherhood. Italy was the place where a 62-year-old woman, Rosanna Della Corte, gave birth to a baby after a donor's egg was fertilized with her husband's sperm and implanted in her uterus. Italy was the place where baby Elizabetta was born—the fetus having been carried by her father's sister from his wife's egg—two years after the wife (Elizabetta's "real" mother) had died in a traffic accident.

In 1995 the Italian national medical association decided that was enough—indeed, a good deal more than enough—and issued a new code of ethics for doctors. It forbade, among other things, artificial insemination for postmenopausal women, artificial insemination after the death of a partner, all forms of surrogate motherhood, artificial insemination for single women, and any selection of sperm based on "the social, economic, or professional standing of the donor."[14] The association urged the Italian Parliament to enact the code into law. The Vatican thought the code was too timid, and proclaimed its own position that *all* artificial insemination is immoral and forbidden. On the other side of the debate, a leading fertility specialist (Severino Antinori, the man who had arranged the 62-year-old woman's pregnancy) called the code "anachronistic, illiberal and anti-democratic" and declared that he intended to defy it. An alliance of homosexuals called it "dangerous and illegitimate" and urged its members to fight it in court. A newspaper report concluded that the doctors' code might curb some excesses, but that the church's total veto would probably "exert no more respect among would-be parents than its prohibition of artificial birth control among the general population."[15] And so it goes in Italy. Elsewhere, I read a prediction that it will soon become possible for a woman to become pregnant at any age—no upper limit whatever. The same source thinks average life expectancy will soon be up around ninety, which makes this prospect all the more striking.[16]

Even when reproductive techniques are not outlawed, they are usually regulated in one way or another, either by government agencies or quasi-official accreditation organizations. Health concerns—such as the possibility of AIDS transmission through donated sperm—make this necessary.

Although these subjects produce all kinds of political controversy —not only about the ethical questions of what should or should not be permitted, but also about the economic questions of who has access to new technological options, and the governmental questions about who gets to decide—our societies and institutional systems are proving to be remarkably innovative in their capacity to manage new technologies. Probably the biggest test of our ability to handle rapid cultural-biological evolutionary change will be the matter of eugenics.

Eugenics Now

In 1993 *The New York Times* ran a story, rich with historical irony, about a community of Orthodox Jews who have a program of genetic testing for young people. The goal of the program is simple: to eliminate common inherited diseases such as Tay-Sachs and cystic fibrosis. Among Ashkenazi Jews, one person in twenty-five is a carrier of the Tay-Sachs gene, and one person in twenty-five is a carrier of the cystic fibrosis gene. When people with those genes marry, there is a one-in-four possibility in each pregnancy that a child will be born with the disease. Tay-Sachs is an incurable, fatal disease in which the child eventually becomes blind and paralyzed. Individuals with cystic fibrosis must cope with lifelong breathing and digestive problems, and about half of them suffer a shortened life span.

So, every year, representatives of the Committee for Prevention of Jewish Genetic Diseases go to the Orthodox high schools and offer the students a blood test. Those tested are given an identification number, which is registered at the program's central office. When a boy and a girl are being considered by the community's matchmakers as likely prospects to be united in marriage, the next step is to call the office hotline with their identification numbers. The office then reports either that the match is compatible, or that the young people both carry a recessive gene and would be likely to produce children with one of the diseases. The ancient Jewish tradition of matchmaking, in short, has moved into the bio-information age.

The members of the community were apparently quite satisfied with the program, which the religious leaders had named *Dor Yeshorim,* Hebrew for "the generation of the righteous," and the practical results were impressive: "Today," a report states, "with Dor Yeshorim's continual testing, new cases of Tay-Sachs have been virtually eliminated from our community."[17] The program was being expanded to test for several other diseases, including cystic fibrosis. But several outside ethicists interviewed by the *Times* were quite worried about it—because, by any name, it is eugenics.[18]

"Eugenics" is one of the most fearsome, explosive words in the whole bioethical dialogue. The media frequently carry reports on research leading toward germline therapy—treatment that would alter an individual's genome and thus the traits that he or she would pass on to future generations—which borders on the subject of eugenics. Sometimes the e-word is actually used, and it is always with the strong implication that eugenics must not be permitted to happen. These statements are quite understandable and their sentiments are commendable, but they are also misleading. They give the impression that eugenics went away—which it didn't—or that it can be excluded from the future—which it can't. Eugenics is part of life in our time, and the challenge is to understand that and manage it wisely.

The word itself—from a Greek root meaning "well born" or "of good heredity"—was coined in 1883 by Francis Galton, Charles Darwin's brilliant but somewhat erratic cousin. Galton proposed to take evolution to its next logical step, from theory to practice, and began to design a rather air-headed program for breeding a superior race of human beings.

Eugenics became a wildly popular cause, as trendy in its time as environmental protection is today. It was championed with equal enthusiasm by right-wing social Darwinists and left-wing socialists such as George Bernard Shaw. It spread to the United States, where generous donors funded research centers such as the Race Betterment Foundation at Battle Creek, Michigan.

Eugenics went berserk in the United States long before it was taken up by the Nazis in Germany. It fed on American enthusiasm for progress, and whispered darkly to nativist fears that the good old American gene pool would be contaminated by the new immigrants. The result was a burst of programs for compulsory sterilization of criminals and mental patients, restrictions on immigration, and laws prohibiting interracial marriage. Many of those programs are now ended, the laws repealed, the research institutes gone. One of the reasons for their demise was the gradual realization that their science and mathematics were a bit shaky. We no longer believe there is any clearly identifiable genetic deficiency such as "feeblemindedness" that is caused by a single gene. Statisticians have calculated that even if such a defect could be clearly identified and all its carriers prevented from breeding (as some eugenic enthusiasts proposed), it would take over 8,000 years to get their numbers down to 1 in 100,000.[19] Then there was the unimpressive example of Nueva Germania, a colony established in Paraguay by Elizabeth Nietzsche (sister of Friedrich). She populated her eugenic utopia with splendid specimens, chosen for the "German purity of their blood," and encouraged their selective breeding toward a race of supermen. The results—still visible in that area—are blond and blue-eyed Paraguayans, most of them poor, inbred, and diseased.[20] For many reasons, the idea of improving the human gene pool lost its momentum as a popular political movement.

But today, as genetic information becomes more accessible, more and more people make decisions that are *de facto* eugenics—whenever a couple chooses to abort a defective fetus and try again, whenever a prospective parent makes a reproductive decision on the basis of knowledge that he or she carries genes for an inheritable disease, whenever a sperm bank screens prospective donors to find what traits they carry. If eugenics is about people-breeding, about attempting to improve the genetic heritage of those yet unborn, all these meet the definition. There may well be more real eugenics going on today than when it was popular.

Most *de facto* eugenics is either short-term eugenics, as when people make choices about which pregnancies to terminate or carry to term, or small-scale eugenics, as in the case of the Jewish community in New York. You can also find implicit eugenic considerations in various policies of public and quasi-public institutions. When an official in the state of California ruled that screening for one genetic defect should be offered to all pregnant women, it was with the expressed hope "that some of those who are found to have children with neural tube defects will choose not to bring them to term."[21] When the 1990 guidelines of the International Huntington Association were written, they declared that it was acceptable to refuse to test women who might be carriers of Huntington's disease unless they gave "complete assurance that they will terminate a pregnancy where there is an increased risk."[22] The sort of nation-sized breeding program that Galton and his colleagues hoped for, that Adolf Hitler and the Third Reich began to put in place—and that most people have in mind when they use the word "eugenics" —is nowhere in sight, but eugenic issues of various kinds are and will be constantly emerging in the bio-information era. As a leading geneticist, Steve Jones of University College, London, puts it: "No serious scientist now has the slightest interest in producing a genetically planned society. But the explosion in genetics means that we are soon

—like it or not—bound to be faced with moral problems about whether we should make conscious decisions about human evolution."[23]

One good argument for thinking open-mindedly about *de facto* eugenics is the constant reality of *de facto* dysgenics—deteriorations of the human gene pool as a result of various social and medical interventions. Anything that medical science does to prolong the life into reproductive years of a person who is born with a genetic illness results in offspring who carry the genes for that illness. As treatments for people with cystic fibrosis improve, and as genetic therapy saves children with severe combined immunity disorder, more of those people will be able to marry and lead normal reproductive lives, and more children will be born with those genes. One prominent medical ethicist, Paul Silverman of the University of California, Irvine, warns that we may be creating a population increasingly dependent on medical care:

> Modern medicine uses a variety of treatment modalities that enable many people to survive to old age who might otherwise have died in childhood. Vaccines and antibiotics protect against a broad spectrum of what previously were prevalent diseases: polio, scarlet fever, tetanus, whooping cough, etc. Congenital malformations are surgically repaired; diabetics and hemophiliacs are injected with essential compounds that they are genetically unable to produce; childhood cancer and leukemia is treated with chemicals, irradiation and bone marrow transplantation. By protecting and treating children to prevent the natural selection effects of disease and genetic deficiencies, we have created a gene pool increasingly susceptible to infectious and neoplastic disease in later life. Without the constancy of public and private hygiene and the use of anti-infection agents, epidemics of unimagined proportions might occur.[24]

Curiously enough, one of the greatest forces for *improvement* of the human gene pool at the present time may be global migration —the very thing that was most feared by the racist eugenicists who believed immigration and interbreeding were increasing genetic defects in the American population. Jones believes, however, that these are producing a significant *decrease* in inherited genetic defects. "Wherever we look," he says, "one thing is clear: there has been a drop in inbreeding in human populations in the recent evolutionary past. An increase in mating outside the group is one of the most dramatic changes in recent evolutionary history. Its effects may outweigh anything that medical genetics is likely to be able to do."[25]

Since eugenic considerations were used to justify anti-immigration legislation in the past, one might reasonably argue now that the present information requires governments to eradicate all boundaries and do everything possible to increase immigration and general moving around in the service of the world's genetic well-being. I don't suppose this is a likely near-term prospect—but Jones's ideas on the eugenic effects of migration, a complete reversal of the view that once dominated Western thinking, does offer an excellent example of how radically perspectives can change in a fairly short period of time. We still have plenty of nativism, of course, but it no longer has its scientific rationale.

I suspect, however, that Jones is wrong in his guess that migration will continue to be a more potent eugenic force than medical genetics. This is undoubtedly true at the present time, but there is reason to believe that the somatic-cell stage of genetic therapy will progress rather quickly into the germline stage—that is, from genetic therapy that treats or cures a disease to genetic therapy that modifies the patient's reproductive genes so that the disease is not passed on to his descendants. I use the "his" deliberately here, because women are born with all the eggs they will use in their lifetimes. But in men, the sperm-producing stem cells are active throughout the reproductive years. Consequently it seems likely that the first germline therapy will be with a male patient. It's entirely possible that this will happen inadvertently, since some of the vectors being used to get curative genes into a patient's cells—such as the adenovirus (flu) and herpes simplex—are known to attack reproductive tissue. In other words, continued treatment of a person suffering from cystic fibrosis—aimed at getting his cells to produce the protein that relieves the typical wheezing and pneumonia—could also get the gene capable of producing that same protein into his reproductive cells, with the result that his offspring will *not* be born with cystic fibrosis. This is a very real possibility which, like so much of gene therapy, has both inspiring and frightening aspects. As of this writing, the Recombinant Advisory Committee is establishing a working group to study the "germline question."[26]

Meanwhile, several deliberate approaches to germline therapy are being tested. One way is to modify the genes of a preembryo during *in vitro* fertilization, before the embryo is implanted in the mother. Another way—already successful with animals—is to insert genes into the fetus. Yet another is to cultivate and alter male sperm, which can then be used for *in vitro* fertilization or artificial insemination.[27] The current prediction is that somatic-cell gene therapy will be standard medicine by the turn of the century.[28] Germline therapy may not be far behind—may not, in fact, be behind at all.

This is, without question, a serious evolutionary step, and one that will no doubt be agonizingly debated as it becomes more apparent that people are about to make it, or are making it—or have already made it. And eugenics will be a part of that debate.

Evolutionary considerations could be used now, in fact, to make a case *against* somatic-cell gene therapy, echoing the case that Spencer and some of the Darwinian true believers made against vaccination: that it would enable people who might otherwise have perished to survive instead and reproduce. Germline therapy, on the other hand, is eugenic.

I doubt that considerations of impacts on the whole human gene pool are going to be the deciding factor for people who are actually making the genetic decisions. But it seems entirely likely that some people whose families have had a long history of producing children with a genetic illness such as Huntington's disease or Tay-Sachs —histories that include enormous suffering—will choose germline therapy if it promises to eliminate disease from future generations. It is quite thinkable, too, that eventually some of those diseases may —through a combination of genetic screening and gene therapy—cease to be of any importance, and go the way of smallpox.

Those are real eugenic possibilities, and as time progresses they will undoubtedly become real eugenic issues that will be debated on economic, ethical, and other grounds. They have to do with a limited —yet large and important—class of disorders that have been established as genetic in origin, usually involving a single gene. They do not concern eliminating vaguely defined forms of inferiority such as "feeble-mindedness," and they do not concern breeding a race of giants and/or geniuses. They do concern real and concrete problems that people are dealing with already, and may soon be dealing with in new ways. The subject of human genetic evolution—and of human actions that may shape its course—is no longer the exclusive province of utopians and racists, although undoubtedly both those constituencies will continue to be heard in the dialogue. Information about genetics is now becoming a part of daily life. And as it does, ordinary people begin to understand that many personal and political decisions have genetic consequences. Such understanding is yet another dimension of the new connections and social obligations that the present evolutionary transition creates. If this is eugenics, let us make the most of it.

PART THREE

THE EVOLUTION OF AGRICULTURE AND INDUSTRY

CHAPTER EIGHT

Reinventing Agriculture

Today, with the advent of modern scientific methods, and with the technologies that accompanied the historical development of those methods, agriculture is indeed a scientific enterprise. From the fundamental understanding of the nature of plants, animals, soils, environments and institutions (especially, markets) to the application of those understandings to food production and processing, science predominates.

—Lawrence Busch[1]

According to recent work by archaeologists at Yale, agriculture was invented about 10,000 years ago by people who lived in the Judean hills at the north end of the Dead Sea. The Natufian culture, as these people are called, were an advanced civilization who had inhabited that region for a long time, and had already established a settled life. They had well-built houses and a sophisticated social structure, and tools such as flint sickles and stone mortars and pestles that they used to harvest and process grains. Then at a certain point they made the transition—one of the more momentous transitions in the entire course of human evolution—to planting grain and cultivating it.[2]

The archaeologists theorize that the innovation was the result of a "convergence of accidents." Four elements had to be present at the same time: genetic resources, technology, social organization, and need.

The genetic resources were available wild grains, which the people were already accustomed to gathering. The technology was their knowledge of how to harvest and process such grains, and the tools they had developed for that purpose. The social organization was essential: a more primitive nomadic society would not have been able to organize the labor and the food distribution. And need: according to this theory, climate had been changing in the Jordan Valley for some time—becoming much hotter and drier. Small lakes dried up, forcing a retreat to the larger lakes in the region, and the population shifts caused food shortages. At the same time, summer droughts reduced the habitats of wild game and shortened the growing seasons of the grains.

But that crisis also presented an opportunity. The warmer climate favored the annual species of wild grains and legumes—those that completed their life cycles in the late spring—over the perennials. The annuals had large seeds, protected inside husks, that were able to survive the powerful summer droughts and then germinate in the cool and rainy winters. Some clever Natufians presumably observed this process and began to help it along a little bit each year by saving seeds when they harvested grains and then planting them in the next wet season. And so commenced—probably at first scarcely noticed—the life of agriculture, with times to sow and times to reap.

It may not have happened precisely as I have described above —and it probably happened at other times and places as well—but it happened. The Jordanian region became agricultural, the people who lived there became a different kind of people. Such shifts from hunter-gatherer life to agriculture were leaps in human cultural evolution that would set in motion an ongoing series of further changes. Food supplies would grow, and populations would increase. Cities would be built, and new religious and political systems would emerge. Changes in genetic evolution also resulted. Without any concept of plant breeding at first, the primitive farmers tended to gather mutant varieties that had tough connections between stalk and seed, making it possible to harvest the seeds and get them back to the village. The archaeological evidence indicates that within a short period of time the cultivated fields in the Jordanian region were taken over completely by the seed-retaining, fat-grained mutants. As agriculture developed there, it spread northward—information leaks, people move—and soon wheat, barley, peas, and beans were being grown in Turkey and Mesopotamia, with corresponding impacts on ecosystems and on the evolutionary careers of various species of plants and animals.

Since then, agriculture has been invented and reinvented many times over. Three major reinventions have occurred in the fairly recent past. The first was scientific plant breeding, a child of the twentieth century. There was no "breeding" at all until the sexual nature of plants became an accepted fact in the mid-eighteenth century. Before then the main means of improving plant lines had been merely selecting and growing the most desirable specimens. With the emergence of genetic science at around the beginning of this century, intentional breeding soon became a worldwide activity, much of it centered in publicly supported institutions. Agricultural machinery was the second reinvention—actually a whole series of inventions. My early recollections of life on a Nevada cattle ranch include vivid images of our hay-mowing machines, which were elegant improvements over the hand scythes of previous centuries—but drawn by teams of work-horses. The workhorses are obsolete now, too, and the stereotypic image of the American farmer has changed—from the man with a plow to the person on a tractor. Mechanization tended to reduce the labor put into food production and increase the fossil fuel input. Then, in the post–World War II years, the development of organic chemicals transformed agriculture a third time. Chemical fertilizers, pesticides, and herbicides greatly increased productivity, but caused major problems: environmental pollution and harm to the health of farm workers. The challenge today is to increase productivity yet again, but to do so this time without creating so many adverse side-effects—if possible, to obtain environmental and health bonuses by reducing fossil fuel inputs and making most agricultural chemicals obsolete.

The Next Transition

The present world situation is vastly different from the local crisis the Natufians confronted, but in some deeper ways it is quite similar. The same four elements—genetic resources, technology, social organization, and need—are converging again. This time the resources are a global library of genetic information that can be adapted to new uses. New technologies—especially biotechnologies—are making it possible to do things with plants that people have never done before. We have a social structure of public and private organizations that, although far short of what the situation really requires, does support some research and development and helps some of the people who produce food get access to new technologies and information. The need for better food production and distribution—born of population growth combined with environmental stresses—is already evident and likely to become more so.

But we have learned, from 10,000 years of history, that major reinventions of agriculture are not just little technological adjustments that take place down on the farm without having much effect on the rest of us. They are steps along the road of cultural and genetic evolution—large-scale systemic transitions that reshape societies, governments, economies, ecosystems. Most of them have uneven consequences—good for some, bad for others—and some have costly and destructive side-effects. Consequently, when another agricultural revolution is seen to be coming down the road, as one certainly is now, different people and groups react with vastly different degrees of enthusiasm—and offer vastly different predictions about what the consequences will be.

Agriculture has its own ideological camps, more or less corresponding to the medical camps. On one hand is the agribusiness establishment, with its heavy machinery, its barrage of chemicals, and its network of old-boy connections among universities, businesses, and government. Its members tend to be rather defensive about the shortcomings of past innovations, and endlessly optimistic about the bright promise of innovations yet to come. On the other hand are various latter-day Jeffersonians who admire—often from a comfortable distance, I have noticed—the joys and virtues inherent in farm life, particularly small-farm life and particularly where it is restrained in its applications of technology. This group includes the organic-farming constituency—who have their own network of food stores, restaurants, publications, and research centers—and many environmentalists. As strongly as the agribusiness establishment clings to the ideology of progress, people of the latter persuasion cling to the conviction that technology is the enemy of humanity and nature, and that technological solutions only create new problems.

Both sides recognize the importance of information in agriculture, but they tend to have strikingly different ideas about what good information is and where it comes from. The farm establishment looks to the laboratories, the agricultural schools, the big producers of seeds and chemicals. For them, the best information is in the form of scientific facts. For the alternative groups, the best information is more likely to be called wisdom, and they are more likely to believe it is found in traditional cultures, or in the knowledge of the practical farmer who gets his or her hands dirty. The farm establishment tends to believe that new science and technology comes in response to needs; the alternative groups tend to believe that new science and technologies are often forced on farmers whether needed or not.

This is more a spectrum than a completely polarized set of distinctly separated groups, and you can find some kinds of farming that creatively stake out a less ideologically rigid territory, much as biofeedback has managed to do in the world of health. Integrated pest management (IPM for short) is the best example of what I am trying to get at here—a sophisticated approach that combines high-tech farming with ecological responsibility. It employs a whole battery of techniques, including computer analysis of insect and predator life patterns, crop rotation, fancy electrical bug traps and—where appropriate—moderate use of chemical pesticides. A farmer using IPM may—instead of getting his pest-management advice from a local representative of a chemical company—hire a consultant, often called a "scout." The scout is a sort of census taker who goes into the field with a device that looks like an oversized vacuum cleaner and collects samples of insects to determine the actual nature of the field's bug population. Analyzing the data, he is able to recommend to the farmer an optimum strategy for controlling pests. This may involve the use of biological controls—other insects —or it may involve applications of pesticide at a certain point in the growing season. The aim is to use the best technique available, to use it selectively and economically, and to avoid the saturation spraying of agricultural chemicals that is, unfortunately, characteristic of a lot of modern agriculture.

That philosophical approach—that creative willingness to put high technologies to work in service of efficient and environmentally responsible production—is the best hope for resolving some of the ideological conflict, and I think it is a preview of the mainstream agriculture of the future. But we can expect many battles between technophiles and technophobes along the way.

Genetic Agriculture

There can really be no doubt, whatever your aesthetic or ideological bias about agriculture, that the links between food production and science have been growing ever closer. All around the world, agriculture is being transformed by scientific discoveries and technological innovations. As a result of scientific plant breeding we have many items available in the supermarket produce departments—and, for that matter, in the organic grocery stores—that didn't exist a century ago. The nature of farming is changing as well—and a lot of it is beginning to look more like what is sometimes called "agribusiness" or "agro-industry."

Consider, for example, the impact made by the work of a couple of breeders in the United States who did some interesting experiments from 1907 to 1910 with corn. They would carefully inbreed certain lines of corn and then cross these strains to produce hybrids—tremendously vigorous and productive plants, but sterile. These experiments were carried forth into systematic field testing in the 1920s, and the hybrids were quickly accepted by farmers when they finally became commercially available. In 1933 less than one percent of Iowa corn acreage was devoted to hybrids; ten years later, it was 99 percent. Farmers were happy with the new corn because of its uniformity, high yields, and disease resistance. Seed companies were happy with it because the farmers had to buy new seed every year. This development produced better corn and more of it, but it also had a downside—the genetic uniformity of the crops. Although there were a number of brand-name maize varieties, they were genetically similar—and similarly vulnerable when a corn blight struck the United States in 1970, and destroyed about half the harvest in several southern states.[3]

Consider one more example: the most spectacular single development in the history of human food production. The Green Revolution produced new high-yielding varieties of wheat, corn, and rice, which resulted in dramatic increases in food production. The conversion was even greater than the one that had taken place when hybrid corn came along. Between 1960 and 1970 over 70 percent of the wheat acreage in Bangladesh, India, Nepal, and Pakistan was shifted to the new wheat varieties. The massive famines that had been expected in India were averted. India, in fact, became a food-exporting country. So did Indonesia, which had once been the world's largest rice importer.[4] But the high yields required ample water supplies and generous inputs of fertilizer and pesticide to perform optimally.[5] Agricultural systems changed, and social, economic, and secondary technological changes followed. Poor people got food, but some big landowners got rich. Bullock carts were abandoned in favor of tractors. "The new technology," one critic reported, "has led to changes in crop pattern and in methods of production. It has accelerated the development of a market oriented, capitalist agriculture. It has hastened the demise of subsistence oriented, peasant farming."[6] Consequently, many people still regard the Green Revolution with disapproval bordering on contempt, see it as something aggressively forced on the public by scientists.[7]

These pieces of history, and others like them, hover in the background of the new scientific-agricultural revolution that is just getting under way. They serve as hopeful proofs that agricultural science can do wonderful things, and as cautionary reminders that wonderful things usually carry a price tag.

For a number of reasons, agricultural applications of biotechnology lagged far behind medical applications at first. One reason was that there was less of a research base to build upon, medical biotechnology having gained a great deal from the scientific legacy of the "war on cancer" that had poured millions of dollars of public funds into research. Another reason was that antibiotechnology activists were able to mobilize public opinion much more effectively against proposed field tests that might have unforeseen ecological consequences, and against new food products that might contain unpleasant nutritional surprises. The first attempts to field-test a biotech farm product in the United States—a genetically modified variety of the *pseudomonas* bacteria (the original variety caused crop damage by forming ice crystals on plant leaves)

—ran into lawsuits, demonstrations, and vandalism in several different California regions where the tests were contemplated. The *pseudomonas* in question was a product of genetic technology; in this case with nothing added, but the ice-nucleating gene removed. The scientists' idea—one of the basic strategies of biological control—was that it would be sprayed onto the plants and would take over the space otherwise occupied by its ice-nucleating cousins. When finally tested, it performed as advertised with no adverse results, but not before the public had been treated to some frightening science-fictionish scenarios of bugs on the rampage if the modified bacteria should migrate and increase uncontrollably. An article in the environmental publication *Earth Island Journal*—uncritically quoting a scenario created by the ever-imaginative biotech-basher Jeremy Rifkin—warned: "By reducing the freezing level of rain falling over major mountain ranges, 'Ice-Minus' could significantly reduce snowfall. Not only would ski resorts suffer the impact of reduced snowpacks but, even more troubling, increased runoff could trigger massive flooding of hills and lowlands."[8]

Another much-publicized landmark of the move into biotechnologically augmented agriculture was the introduction of the Flavr Savr tomato. This was actually the third product of recombinant DNA technology to receive approval from the U.S. Food and Drug Administration. The first, in March of 1990, was Chy-Max, the biotech equivalent of the enzyme from calves' stomachs that is used in cheese manufacture. For some reason nobody paid much attention to Chy-Max, and within a few years 50 percent of processed cheese products were derived from that source. The second product was bovine somatotropin (BST), a growth hormone that stimulates milk production in cows.

The first use of BST led to a period of heated and confused controversy. Antibiotech activists formed common cause with some small milk farmers, who feared they would be driven out of business if the big dairies became more productive. Questions were raised about possible harm to consumers from BST in the milk, and this led to much argument about whether there was any difference between the recombinant BST and the BST that all cows have in their milk anyway. The politically correct ice cream company Ben & Jerry's began labeling its products with a declaration that the firm did not buy milk products from dairies that injected cows with BST. This was good enough for some consumers, although other critics grumpily pointed out that the company's "super premium" ice cream might be free of unnatural BST, but was larded with natural—and lethal—milk fat.[9]

Then along came the Flavr Savr tomato, product of antisense technology, designed to ripen on the vine but *not* to turn into scarlet mush before getting to the table. This was the first product that consumers could bite into knowing they were eating a product of biotechnology, and it occasioned a great deal of controversy—the main question being whether it was somehow nutritionally deficient or even dangerous. I attended a number of conferences at which these issues were debated, and the only convincing anti-Flavr Savr argument I heard was the suggestion that some people might be allergic to the product. This question has still not been fully answered, but the most recent reports indicate that, after being delayed, the tomato is finding its way into general acceptance and use. The delays had more to do with production and distribution problems than with any of the heady issues that were discussed at conferences and cited in the many newspaper articles about unnatural tomatoes. There is a connection between the tomato's career and that of BST, in that Monsanto—the BST manufacturer—was so pleased by the strong sales of its product after the initial storms had passed that it decided to make an even heavier move into agricultural biotechnology. It proceeded to acquire a portion of Calgene, the tomato company, and to set up a new arrangement that would make Calgene a major packer and shipper of fresh tomatoes.[10]

Although agricultural biotechnology was a bit slow getting out of the gate, it is now rapidly proceeding on its revolutionary way. We may look forward to new cooking oils without saturated fats, fresh-picked-tasting corn two weeks off the vine, decaf coffee from decaf coffee plants.[11] The revolution is taking place not only in the United States, but around the world. Not long ago I attended yet another conference, this one a gathering of international food biotechnology researchers —scientists from Africa, Asia, South America, Europe. One of them confidently predicted that all major food crop plants will have been transformed by the end of the century. This doesn't mean that all crop plants grown everywhere will be the products of biotechnology—only that some varieties of each crop will have been reengineered. The conferees reported progress on modifications of basic crops such as yams, tomatoes, bananas, wheat, and corn. Most of the early modifications are in the direction of pest and disease resistance, higher yields, and improved nutritional qualities. A bit later will come nitrogen fixation—eliminating or reducing the need for chemical fertilizers—and plant variations with higher salt or heat tolerance.

Some of the changes resulting from these efforts won't be particularly dramatic—or even noticeable. Many people will go on growing the same crops—frequently in the same way—even though they have been genetically transformed. There is really nothing different in the way you grow a virus-resistant potato if you were too poor to use pesticides in the first place; all that happens is that you get more edible potatoes.

But in other cases, of course, farmers will switch to new commercial crops, with many social, economic, and ecological consequences. Agriculture will change, and in some places it will change with dizzying speed and on a vast scale. One reason it will change is that biotechnology is not the only transformative force looming over the farm horizon. It is merely one part of the new "informatization" of agriculture.

The Knowledge Farmer

Information has always been a central part of agriculture, and its role is becoming increasingly important. Farmers will have to learn new things, and more and more of them are likely to do their work in collaboration with smart machines or with advanced information-support systems—or both.

Here we have a Minnesotan out driving his tractor across the rich fields of his 700-acre farm. The tractor itself was a revolutionary change when it came along, but by now, of course, its presence is not going to knock your socks off. What you want to pay attention to here are the radio receiver and the laptop computer mounted in the farmer's cab. The computer contains a software program to manage the distribution of chemical nutrients and data on the composition of the soil in each part of the field. The radio receiver picks up satellite messages—relayed from a nearby tower—that pinpoint the tractor's exact location and feed further information to the computer. The computer determines how much fertilizer is needed for each fraction of an acre, and doles out precisely that amount as the tractor rolls across the flat Minnesota fields. This is called "precision farming," and is also referred to in some places as "site-specific farming," "satellite farming," or "prescription farming." Although it may sound like a computer nerd's fantasy of country life, it is proving to be extremely effective in cutting down on the use of fertilizers and pesticides—with environmental payoffs as well, since excess fertilizer normally leaches into ground water.[12]

You might expect to find such high-tech ventures in the United States and other advanced industrial countries—there is a real and serious "information gap" between such countries and the less developed nations—but the reinvention of agriculture now under way is based essentially on bio-information, and thus has the ability to spread very quickly to other regions. All around the world you can find projects dedicated to leapfrogging over the heavy-machinery and heavy-chemical stages and proceeding directly toward new farming activities that harness information technology and biotechnology to the service of small-scale sustainable agriculture.

In India, for example, a number of such projects are guided by the Swaminathan Foundation's Center for Research on Sustainable Agriculture and Rural Development. Some of the activities are not conspicuously high-tech: restoring the fertility of saline soil in coastal areas, developing more effective irrigation systems, trying some new crops in agroforestry. But the program is also helping local communities build up their own gene banks of farmer-conserved and farmer-developed plant varieties, and the foundation's informatics center holds a collection of computers and an electronic library of CD-ROMs. The center has connections to the Internet, and thus to a worldwide linkage of information services. It is even powered by a solar voltaic generator.[13]

International networks of such activities have grown up, and so has a body of research around what is now called "technology blending." This is an approach to development based on the assumption that scientific and technological progress will inevitably impact small farmers, and that—instead of waiting for it to come crashing down from above—we might try to help people find new and more productive ways of doing things without massive disruptions. So the Minnesota farmer is not entirely alone out there on the bio-information frontier. Interesting and surprising things are going on all around the world. In Egypt, on some of the world's oldest irrigated farmlands, lasers are being used to level land for optimum water utilization. In Burkina Faso satellites help map out routes for rural roads. In Malaysia a computerized information network serves the needs of small rubber farmers.[14]

Biotechnology, then, is merely one element in a convergence of developments in machinery, information systems, global trade, and agricultural science that are carrying us quickly into another chapter of the story that may have begun in the Judean hills. "Horse plows came and went," says a member of Purdue University's agriculture faculty. "Steam engine machines came and went. Then the farm tractor came, and now the only place it's going is to the antique museum."[15]

Food without Farms, Farms without Food

While agriculture evolves, entirely new ways of producing food appear—ways that are scarcely agriculture at all. Something on the other side of agriculture is in the process of being invented. And a new kind of agriculture appears also—producing not food, but medicines.

In 1990 I wrote an article predicting that it would someday become possible to produce "real" food items—such as fruit juices, flour, vegetable oils, jellies, and tomato paste—in factories. My research at that time drew mainly on the work of an interesting odd couple of scientists:

Martin Rogoff, a microbiologist who was then director of the Department of Agriculture's research center in Albany, California—not far from where I live—and Steven Rawlins, a soil scientist at the Systems Research Laboratory in Beltsville, Maryland. The federal government had put the two of them together to work on a long-range strategic planning project, of a sort that you would be more likely to encounter in some secret defense agency than in the rather staid corridors of the USDA. Their charge was to think about what might happen if some event should cause a fundamental disturbance in America's food-production system—anything from a major cutoff of fossil-fuel supplies to the desertification of the "breadbasket" regions by soil erosion and/or climate change. The charge was to think about, as Rogoff put it, "what would happen if we really got whacked"—and, of course, to think about how the worst effects of such a disaster might be avoided.

The first thing that became disturbingly apparent to the two was the extreme vulnerability of the American food system, its shortage of backup supplies. We have stores of some food supplies such as grains, but scarcely any reserve supplies at all of most of the other things that people eat—such as vegetables. While Rogoff and Rawlins worried about this, they also thought about what was going on in biotechnology research, and became excited about the idea of an alternative food-production system, based on tissue culture, that might be developed to serve as an emergency backup.

Tissue culture is one of the dark horses of biotechnology—less famous and less controversial than gene splicing, simpler in principle but no less impressive in results. Plant scientists can take a tiny slice from the leaf of a tree and, by cultivating it in a medium of hormones and nutrients, grow it into a whole tree. This process can produce a hundred, or for that matter a million, copies of a single tree. Another kind of tissue culture involves growing only a certain part of the plant—such as the edible fruit.

After I had become acquainted with the work of Rogoff and Rawlins I took a trip down to Southern California to visit a man who was doing tissue culture research with fruit. His name was Brent Tisserat, and he had a laboratory in Pasadena where he worked on growing the juice vesicles of oranges, grapefruit, lemon, and other citrus plants—just the juice vesicles, the small glandlike organs within the fruit. In these experiments he was developing the right growth medium of chemicals that would instruct the cells to develop into juice sacs.

From those tissues came juice—real orange juice, but produced without oranges or orange trees. Tisserat was *not* doing this work with a dream of producing some *Star Trek* food system. Mainly he was trying to benefit conventional farmers—the people out in the hot citrus groves not far from Pasadena who were growing oranges and grapefruit—by finding new ways to test the effects on their crops of different nutrients and pollutants. He was not really enthusiastic about the reporters who periodically heard about his work and came around to see if he could really squeeze juice out of a test tube. But the fact of the matter was that he could—albeit in mighty small quantities.

Other people are doing similar tissue-culture experiments. Don Durzan, a pomologist at the University of California at Davis, had been having good luck growing the edible portion of cherries in culture. He was similarly cautious, noting that nobody was yet ready to scale up such a process to the point of producing cherry jelly in a factory.

But producing cherry jelly in a factory—or fruit juice, or a number of other real artificial food products—was precisely what Rogoff and Rawlins thought might be possible. To understand that possibility, you have to think of the tissue culture as a part of a system, something closer to the invention of the wheel than to the invention of the automobile. A working food factory would have to have a steady supply of sugar, which is a basic element in the growth of all plants and all the parts of plants that animals eat. A complete plant such as an orange tree creates this supply through photosynthesis, taking carbon from the atmosphere and combining it with oxygen and hydrogen to form sugar.

Rogoff and Rawlins proposed that feedstock crops—trees or bushes—could be grown on land to which they were well adapted, with minimal inputs of water and agricultural chemicals. They would not be as vulnerable as the California orange groves—which are protected against chills by energy-expensive heating systems, and protected against fruit flies by helicopters spreading pesticide. The crops would not have to be harvested annually, so food energy could be stored in the living plants. (Other forms of biomass, such as wood chips and straw, could also serve as the raw material. There is plenty of biomass in the world, as you may have noticed after a day of gardening or sidewalk sweeping.)

Wood and straw biomass are mostly lignocellulose, tough chains of glucose molecules. Human stomachs are not capable of digesting lignocellulose, but biotechnology researchers are now developing various ways to digest it artificially and turn it into simple sugar syrups. Lignocellulose becomes the missing link between the sun and a fruit-juice factory.

The system, then, would have three main elements. First, the source crops. Second, the conversion plant for turning biomass into sugar syrups; this would probably be located, for maximum efficiency, close to where the trees or brushes were grown. Third, the food-production centers, which would be located close to consumers. Its appeal and logic as a backup system lies in its flexibility. With conventional agriculture, you don't have a lot of room for short-range response. If you have planted the crops, you harvest them. If you have an orange grove, you either pick the fruit when it's ripe or you lose it. The tissue-culture system would be demand-driven, capable of increasing or decreasing production, or switching products, on relatively short notice.

It might also be useful—as many people are now saying—for something more than emergency food production. It might provide part of the food supply for fast-growing cities in Third World countries, or for people living in intensely cold regions where fruits, vegetables, and variety foods are obtainable only at great expense and with heavy inputs of energy. Another advantage of this system, as Rogoff pointed out, is the fact that the feedstock crops would take carbon dioxide out of the atmosphere. Some people are now advocating huge forest plantations to sequester carbon dioxide, and any forest-based agriculture would contribute something to such a strategy.

Of course, you have to be careful about getting too enthusiastic about such things, and in my article I cautioned that, although the research is intriguing, it would be a long time before anybody could scale up any such process for commercial production. "Clearly," I wrote, "nobody is about to put the pieces together this year or next."[16]

Well, I was wrong. In 1991, about a year after I wrote that article, a biotechnology company in California took out a patent on a process for producing vanilla extract by cell culture, and formed a production partnership with Unilever. The product, they said, would be equal in flavor quality to the precious and savory—and extremely expensive—extract of beans from the vanilla orchid, which grows in a few tropical regions such as the island of Madagascar. But it would be far better than the synthetic vanilla now sold in most stores. Cell culture has already begun to compete with agriculture, and a new era in the production of food and fiber has (very quietly) begun.

I don't expect agriculture—either large or small—to be replaced by this method. What the new method offers, rather, is an expansion of ways to produce food, variety food products such as vanilla extract, and possibly fiber: a researcher in Texas is growing cotton fibers in tissue culture. In the global bio-information society, there will be many kinds of agricultural and nonagricultural food production.

Cell culture's appeal lies partly in its potential as an *urban* food production system, a contribution to an important—yet all too often overlooked—part of the world's food supply. Urban residents all around the world now produce food—in backyards and on balconies, in community gardens and plots of available land among the streets and houses. Fruit trees grow along sidewalks and roads; fish destined for the dinner table swim in urban ponds and waterways; city-bred animals are raised for meat, eggs, and milk. Now, with urban populations increasing—while farmland disappears and deteriorates and the environmental and economic costs of developing new farmland become more critical—this part of the food-production system deserves more attention than it usually gets. The UN's Development Programme encourages it, and there are even a few cities—such as Jakarta, Indonesia, and Buenos Aires, Argentina—that provide information and assistance to residents who want to produce their own food. But most municipal governments either ignore such activity, hobble it with restrictions, or prohibit it outright.[17] Here and there you find visionaries such as Nancy and John Todd of the New Alchemy Institute in Massachusetts, whose ideas of urban design include warehouses converted to food production—lettuce on the third floor, hydroponic vegetables on the second floor, chickens and fish on the ground floor and mushrooms in the basement—and sidewalk solar aquaculture systems growing fish in translucent plastic tubes.[18] These ideas are not entirely fanciful; all of them have been tested at the institute, and the Todds raise tilapia and catfish for their own use in plastic tube tanks in the backyard of their Cape Cod home. All of the work of the Todds and New Alchemy Institute is out of the organic/ecological tradition, but other researchers are exploring avenues such as genetically engineered mushrooms and vaccines against the diseases that sometimes infect city-grown fish. Future urban agriculture, like its country cousin, is likely to come in many forms.

The Family Pharm

Someone coined the term *pharming* to describe a new kind of agriculture based on the use of plants or animals genetically modified to produce specialized items for human medical use: goats whose milk contains the heart-attack remedy TPA, sheep that generate a treatment for emphysema, potatoes that carry a human blood protein used in surgery.

Pharming, like older kinds of agriculture, comes in two packages: plant crops and animal husbandry.

Ironically, tobacco plants—not high on any health activist's list of favorites—are particularly suitable for being adapted to producing medicinal products. The plants of the tobacco family have been extensively studied, and were the crop of choice for much of the early research. The first "pharm" product—a sunscreen based on human melanin grown in tobacco plants—is now on the market in Europe and awaiting regulatory approval in the United States. Several writers on the subject of pharming have pointed out the strange but distinctly likely possibility that tobacco farms growing lung-cancer medicine will be a part of America's agricultural future. Other early entrants in this new field of agriculture are potatoes that carry serum albumin—a blood protein used in surgery—and rapeseeds containing a painkilling chemical called enkephalin that occurs naturally in the human brain.

Most of these drug-producing plants are created through a combination of recombinant DNA and conventional breeding: genes are spliced into embryonic plant cells, which are then grown to maturity and repeatedly cross-bred with other similarly modified plants until a new crop line has been established. But Biosource Technologies, a company in California, came up with an intriguingly different approach. They use tobacco as the drug factory, but instead of splicing new genes into the plant, they splice them into the tobacco mosaic virus that infects the plant. The virus then takes over the job of introducing the gene into the plant, causing its cells to produce the protein encoded by the new gene without actually changing the genetic makeup of the plant. The tobacco leaves are then harvested and ground, and the resultant proteins—such as interleukin-2—are extracted and purified. The company's CEO calls this method "geneware"—short for genetic software—and says it takes less time and produces higher yields than the "old-fashioned" recombinant DNA–plant breeding approach.[19]

Then there are transgenic animals, created by the same science that brings us the new varieties of laboratory mice and the pigs with human shield protein. Most of these will be dairy animals, goats or cattle, that give medicinal milk.

This promises to be a major new source of medicinal products, and it has important implications for developing countries—because it is a lot easier and cheaper to raise and milk a herd of transgenic goats than it is to build and maintain the high-tech fermentation processes that are necessary for the current generation of biotechnological medicines. Genzyme, a biotech company in Massachusetts, has for some years been developing goats that produce tissue plasminogen activator (TPA), the much-discussed heart-attack treatment protein, in their milk. TPA has just been given a new vote of confidence by the U.S. Food and Drug Administration, and is in great demand. And, as produced in American factories, it costs $2,200 a dose. There will be economic hurdles to cross before this can become a practical mode of agriculture in developing countries: the prospective medical goatkeeper needs to get the animals first, and they will cost. Issues concerning patents need to be internationally resolved. But the difficulties are surmountable. This is a kind of biotechnology that has a real potential to generate new sources of income for small farmers and businesses.

Pharming may well move—much more quickly than gene therapy or tissue-culture food production—from absolutely nowhere into the mainstream of all our lives. I expect that it will be a recognized and growing component of global agriculture before this century is out.

Undoubtedly there will be some unpleasant ecological and nutritional surprises as agricultural biotechnology unfolds—people do not do things without making mistakes—but in a way the most serious threat of the new agricultural science is that it will fulfill all its promises without too many serious glitches. Because if that happens, the world is in for a stunning change.

Losers and Winners

Let us examine another piece of history—the story of indigo—that may tell us something about the future, and about what might happen if the new agriculture lives up to its expectations.

There was a time when fortunes and colonial empires were built on indigo. It came from certain kinds of shrubs yielding a colorless chemical that, with a little application of very ancient biotechnology —a technique that had been known since before the time of Christ— could be converted into a deep blue dye. As the European textile industry developed, huge plantations in tropical countries were devoted to indigo production—nearly 2 million acres in India. Then, around the beginning of this century, German scientists decoded the chemical structure of indigo—thus making it possible to synthesize it in the laboratory—and factories began manufacturing it. The result was an economic disaster in India, and the birth of a huge new industry in Germany.

Agricultural economists who know what is going on in biotechnology suspect that a modern-day replay of the indigo story is set to happen. It is already happening wherever consumers or food manufacturers choose artificial sweeteners over sugar. It is likely to happen again, many times over, as the industrialized countries perfect new ways to produce goods —such as oils, flavorings, fibers, and still more sweeteners—they once imported. I have read European Community studies about using biotechnology to convert some of Europe's surplus agricultural productivity into a source of chemical feedstock that would lessen the dependence on imported petrochemicals.

How will this affect the countries that now rely heavily on the export of certain crops? What will happen when their customers find new ways to produce the same things at home, and more cheaply? The results will vary widely according to different levels of access to information, and ability to respond. It is entirely possible that some countries will be hit hard by the loss of markets for their major crops. It is equally possible that other countries may anticipate the substitution and—with the right combination of genetic resources, technology, and social organization—creatively develop new cash crops, increase domestic food production, and perhaps even do some ecological restoration on land that has been degraded by decades—if not centuries—of monoculture.

This is one of the areas in which bio-information has the potential to bring about massive economic impacts, and fairly soon. In an information-age world economy, with research and development proceeding at a rapid clip, the production of food and fiber (and pharmaceuticals) could become almost as mobile and changeable as some kinds of manufacturing and services have become in the recent past. Some agricultural regions that were traditionally associated with certain crops may change fundamentally—as indigo-growing regions did in the past and sugar-growing regions are doing now—and change again. As that happens, the value of some natural resources that now underpin national economies will shift, and so will ideas about national interests and security.

Patenting and other protections of intellectual property rights become increasingly thorny issues in this emerging global food-production system. For decades, seed companies and plant breeders have been able to obtain "patentlike" protection under national laws and international agreements, and now it is possible to obtain full patent protection for plants, microorganisms and animals that meet the criteria of "novelty, nonobviousness, and utility." The number of applications for agricultural patents has skyrocketed, and observers of this scene are heatedly debating the legal and ethical pros and cons. Will Third World farmers have to pay American seed companies for patented varieties developed from plants native to their own countries? Will the promise of patent protection stimulate research by smaller companies, or benefit only the big corporations? Will university researchers organize their work around the promise of patentability rather than scientific importance? Does patenting (which requires publication) encourage the free movement of knowledge among scientists, or discourage it?[20] How protectable are plant patents going to be, and how protectable should they be?

Even while patenting plays a larger role in agriculture, researchers in many places are busy developing new plant varieties that will *not* be patented. That was the case with the strains developed for the Green Revolution, and it is one of the arguments in favor of research supported by governments and nonprofit institutions.

The Future(s) of Agriculture

If the whole global food-production system does indeed become more changeable than it has been in the past, this will further accelerate the "informatization" of agriculture. Farmers will need not only more information about how to produce crops, but also more information about markets and social or ecological factors that might affect people's needs and preferences. Farmers—and pharmers—will need to be able to respond to such information, and agricultural science will need to develop sufficient vision and foresight to respond to ecological feedback. The prospect of global climate change is, perhaps, the simplest and most dramatic example of something that may force large-scale reinventions of agriculture. But the shift that I am describing here, the transition into a more flexible, responsive, information-based, highly pluralistic—and globally interconnected—food-production system, will take place in any weather.

CHAPTER NINE

Bio-info Industries

The primacy of mathematical physics as the science of sciences, as the exemplary core of general scientific progress, which it has been since the seventeenth century, is now passing. The new hub is that of the life sciences, of the lines of inquiry that lead outward from biology, molecular chemistry, biochemistry, biogenetics. . . . These lines now seem to radiate and spiral toward every quarter of scientific and philosophical pursuits, as did the physics of Descartes and Newton.

—George Steiner[1]

It is obvious by now that the bio-information era is bringing in a medical revolution—indeed, a veritable crowd of medical revolutions—and it is equally obvious that a massive reinvention of agriculture is also under way. This was to be expected, since medicine and agriculture are both essentially biological activities. What is far more uncertain at this point is how far the bio-informatization process will spread into other areas, particularly industry. Will we see, as some have suggested, a "greening of industry" or "second industrial revolution" in which many kinds of work that were formerly done with mechanical tools and mineral resources begin to be, in one way or another, based on the new biotechnologies—a switch from hardware to geneware?

I think we will. If, as Steiner wrote, the life sciences are now occupying a new role in relation to human thought and progress generally, this extends far beyond the universities and the laboratories. It means a larger restructuring of human civilization, a transformation in which the biological revolution—or more accurately the bio-information revolution, since the convergence of biological science and information/communications technology has now formed a new entity that is no longer separable—leads the way.

As this may sound like a bit of future-hype, I should point out that there is nothing new about making things from animal and plant materials, and that biotechnologically produced enzymes are already used for many industrial purposes and also in household detergents.

Plants are an important source of raw materials for several kinds of commercially manufactured products including soaps, adhesives, inks, detergents, dyes, and pigments—important, yet still modest in comparison to other sources. The Institute for Local Self-Reliance, a Minneapolis think tank that has done extensive research on this subject, reports that fewer than 10 million tons of plant matter (other than wood) are used in industrial and construction products in the United States, in contrast with about 175 million tons of petroleum and coal, and 300 million tons of inorganic materials such as sand, salt, and iron ore. However, the use of plants has been increasing steadily in recent years.

And an impressive number of new bio-industrial processes are now appearing. These aren't always revolutionary jumps from physics to biology—but rather practical integrations of organic materials and biotechnology where it is efficient and commercially viable. This is another case of a disappearing boundary. We noted earlier the fading distinction between biotechnology and conventional biology. Here we see some replacements of the inorganic by the organic, but we also see many mergers of what is biological with what isn't. We also see—in the field of bioremediation—biotechnology cleaning up after heavy industry.

Bioremediation

Frequently, in my routine travels about the San Francisco Bay Area, I observe streams of motorists passing heedlessly by ordinary-looking construction sites where a rather extraordinary industrial operation called bioremediation is taking place. Bioremediation is the use of bacteria to clean up toxic wastes. Nobody pays much attention to these places. Perhaps a few drivers would hit their brakes if somebody posted a "MICROBES AT WORK" sign along the road.

In one such project, a shopping center was built on land that had formerly served as the headquarters for a trucking company. For decades, beginning at a time well before there were any environmental regulations to speak of, oil and gasoline and diesel fuel had soaked into the ground there—and into the water under the ground—from leaking fuel tanks and the general messiness of trucking operations.

The government now requires developers to clean up such sites before they can build anything new on them. One way they can do this is to dredge up all the contaminated soil and haul it away to a toxic waste disposal site. Bioremediation is another way. In the case of the trucking lot, a local firm called CytoCulture pumped the water out of the ground, physically separated out as much of the oil as possible, and then processed the water with oil-eating bacteria until it was clean enough to drink. They also piled up the contaminated soil, inoculated it with bacteria and nutrients to stimulate their growth, then covered it and pumped oxygen through it until the germs had biodegraded the pollutants.

Bioremediation grew out of studies in the late 1960s and early 1970s, mostly supported by the U.S. Office of Naval Research. Prior to the creation of the Environmental Protection Agency, the Navy was in charge of cleaning up the seas, dealing with the marine petroleum spills that were already being recognized as a major global environmental problem.[2] In 1989 bioremediation got its most public challenge along some parts of the Alaskan shoreline that had been fouled by the Exxon Valdez oil spill. It was a fairly simple procedure—scientists working for the U.S. Environmental Protection Agency merely sprayed the beach with fertilizer to stimulate the growth of the local bacteria—but the results were impressive and led to stronger EPA interest in the approach.[3]

Bioremediation of petroleum wastes is now a growing industry around the world, and its promise has inspired a global search for pollutant-hungry bacteria that can be put to work cleaning up petrochemicals and other toxic wastes. Some aspects of this quest have all the scientific sophistication one might expect, and some are surprisingly unfancy. To give an example of the range, researchers from the Woods Hole Oceanographic Institute discovered naphthalene-degrading bacteria near warm water vents 6,000 feet below the surface of the Gulf of Mexico —and another scientist cultured a usable strain of petroleum-eating bacteria from an oil spot he scraped off his driveway.

The culturing usually involves selective breeding—growing the bacteria in solutions of the materials they are intended to clean up, so that strains with really healthy appetites for the specific chemicals are developed. Other people are creating new bacterial strains through recombinant DNA technology; this was what led to the historic *Chakrabarty* decision, regarding the first patented microorganism, a bacterium designed to clean up oil spills. Dr. Ananda Chakrabarty's invention was only a legal success: it established the precedent for patenting life, but was never used in the field and probably would not have been particularly effective. However, plans are now under way for a trial run of a genetically engineered *Pseudomonas fluorescens* bacteria—said to be able to consume various toxic chemicals including naphthalene, anthracene, and phenanthrene—in polluted soil at Oak Ridge National Laboratory.[4] The use of such bacteria may well turn out to be a great forward step in environmental cleanup, and it may also turn out to be the source of new public outcries and media circuses comparable to the one, mentioned in the previous chapter, that attended the first field test of the ice-minus bacteria in California.

But naturally occurring, un-genetically engineered bacteria can be found to work on an amazingly wide range of pollutants—many of which you would not expect to be regarded as edible by any living creature. One strain that was isolated from sediments in the Potomac River converts a soluble form of uranium into an insoluble form—so, if uranium-polluted water is passed through a "bioreactor" containing these microbes, they cause the uranium particles to precipitate out and settle to the bottom, where they can be collected.[5] Other materials that can be decomposed by biological means include polychlorinated biphenyls (PCBs, a major urban pollutant), selenium wastes, and many pesticides.[6] As one bioremediator puts it, "There are bacteria that will eat anything."[7]

I can't verify that statement, but I did see an impressive display in a laboratory that contained a fifty-gallon barrel full of bacteria in a solution of water and phenol. The vat gave off a strong, medicinal odor. Phenol is a powerful chemical: lab technicians use it to disinfect their work surfaces, and dermatologists use it in those wonderful ointments that take off outer layers of your skin. You would naturally expect that the phenol would kill the bacteria—and you would be wrong. The bacteria were eating the phenol, absolutely thriving on it and in the process removing it from the water. In another barrel, connected by a pipe to the first, was pure, clean-smelling water. This was what was left after the bacteria finished with their meal. The phenol was, for all practical purposes, gone—and so were the bacteria, which died off after there was nothing left for them to eat.

Bioremediation is not always that successful, and rarely simple. It is not accomplished by blithely scattering about bacteria or nutrients and waiting for everything to be cleaned up. It is most effective when a number of variables such as temperature and oxygen supply can be controlled. It is, so far, a technology effective in limited ways, its strengths and weaknesses not fully explored. Nobody is calling it the magic solution to all the world's pollution problems. But it is a recognized industry, well beyond the experimental stage now, and one very good example of a new way to apply biology.

Biomining

Using bacteria to recover metals from ore is (like some kinds of bioremediation) based on the interactions between microbes and metals that are part of the Earth's stunning and seemingly endless evolutionary variety. Some kinds of microbes draw energy from metals—feed on them and live on them—and the metals are changed in the process. The changes often make it easier for people to recover the metals in usable quantities. This may seem like a rather exotic and unlikely application of biotechnology—but it has a long history. People in the Mediterranean area were recovering copper by draining water from mines a thousand years before the time of Christ, and the technique of using water to leach metal from ore has been used in many times and places. It was only in the relatively recent past—1957—that scientists learned bacteria played a key part in this process. Like the famous Molière character who realized one day that he had been speaking prose all his life without knowing it, people had been doing biomining without knowing it. Now they are doing it deliberately and a bit more scientifically: over 10 percent of the copper produced in the United States is leached from ores by microorganisms.

The bacteria used in such biomining processes are those found naturally in the ores, where they have evolved. But three forces—the depletion of high-grade ores, the higher costs of energy for mining operations, and the growing base of information about genetics—are likely to increase the use of bacteria in mining. As in bioremediation, some of the microscopic mine workers will be bacteria cultivated (or selectively bred) from naturally occurring varieties, while others will be newly created by recombinant DNA or other high-tech approaches. One researcher, for example, has identified genes of a strain that tolerates arsenic. Since some gold veins are rich in arsenic, this might result in a way to get around the expensive—and polluting—arsenic-removal techniques that are used now.[8]

Biomaterials

While some scientists search through the world's library of genetic information for medicines and foods, others are looking for ways of making new materials for industrial use. Can the genes from oysters produce an underwater glue? Can a spider's genetic code be used to make a polymer that is five to ten times stronger than steel, that can be pulled out to 20 percent of its length without breaking?[9] Neither of these is close to being a commercial reality at the moment, but neither is a possibility that anyone knowledgeable about biotechnology would dismiss out of hand.

Scientists have now managed to clone genes for one of the two proteins that make up spider silk, and expect that it is only a matter of time before it will become possible to manufacture it in more or less the same way that human insulin is manufactured now. One specialist in the chemistry of spider silk is already talking about how it could be used: it might, he thinks, make excellent nonscarring surgical sutures, about one-tenth the diameter of current ones. Or it might go into artificial ligaments and tendons—either as a complete replacement or as a material to mend damage. Military researchers are thinking about its usefulness in parachute strings or bulletproof vests—or possibly for tethering objects to space vehicles. Others speculate about how it might make durable sports equipment, or even replace steel cables in suspension bridges.[10]

That's a bit far out, but it is technically possible now to use biomass feedstocks—as materials made from plants and animals are called—instead of petroleum-based chemicals in quite a wide range of industrial applications. A Department of Energy study concluded in 1994 that nineteen of the fifty most widely used petrochemicals in the United States could be replaced by biomass sources if necessary.[11] These petrochemicals are not being replaced now because they perform well and are more economically practical. But a number of factors, singly or in combination—increasing cost or decreasing availability of petroleum, research breakthroughs that make biomass sources more economical or effective, environmental regulations, pressure from consumer groups—could change that picture.

Some of the hottest prospects for the future of biomaterials are biodegradable plastics. Petroleum-based plastics are extremely unpopular with environmentalists, and many biotech labs have been hard at work looking for polymer products that could be commercially useful yet degradable in the environment.

One product that appears to be making a go of it commercially is Eco-PLA (pronounced *eco-play*), a plastic an American company, Cargill, makes by fermenting corn to produce lactic acid. Polylactic acid, as this product is called, has been manufactured for a long time—and used mainly in surgical products such as sutures that break down in the human body—but used to cost hundreds of dollars a pound. Cargill scientists designed a new process that enables the company to get the price down to around a dollar a pound, and analysts expect to see Eco-PLA widely used in such products as fast food containers and trash bags—and, of course, plates and utensils for picnic meals. The company claims that the material can be recycled easily or can be totally degraded through composting, which turns it into water, carbon dioxide, and the soil nutrient humus, in sixty days or less. This is a considerable advance over the "biodegradable plastics" that appeared amid much fanfare in the 1980s. Those, it turned out, didn't actually biodegrade but only fell apart into minuscule pieces of polyethylene, which were still basically good old plastic. Theoretically, Eco-PLA could end up decomposing in the earth of a cornfield whose crops will ultimately be used to make more plastic. It can be manufactured from a number of grains, potatoes, and sugar beets, but the company that developed it is in the corn-milling business and expects it to provide the biggest boost to the market demand for corn since the widely used sweetener high fructose corn syrup (HFCS) was developed in the 1970s.[12] (HFCS is, by the way, one of the products that has contributed to the declining value of sugar beets and cane.)

Plastics—which are basically polymers, long molecules with repeating sequences—occur in nature in many ways. Back in 1926 French agronomist Maurice Lemoigne discovered that some bacteria use a type of plastic instead of fat to store energy reserves, and subsequent researchers have found more than ninety kinds of bacteria that manufacture plastics. A company in Europe has been using one such, *Alcaligenes eutrophus,* to make another kind of biodegradable plastic, polyhydroxybutyrate. This is an expensive product, but ecologically respectable and used in items such as soft drink bottles and disposable razor handles. More recently, Christopher Somerville, a researcher at Stanford, spliced the plastic-producing gene from *Alcaligenes eutrophus* into a relative of the common mustard plant. The plants grew nicely and made a substance that closely resembles polypropylene, the kind of plastic used to make products like gallon-sized milk jugs. Monsanto purchased American rights to the process, and plans to try it in rapeseed plants. If the process proves effective—performs well in field tests, survives the regulatory hurdles—it may end up being yet another odd crop for farmers of the future, along with sunscreens and heart-attack remedies.[13]

Bio-energy

People have obtained light and heat over the centuries from a wide variety of biological materials, from yak dung to whale oil to plain old firewood, but the industrial revolution ran on nonrenewable fossil fuels: first coal, later petroleum. As the bio-information revolution unfolds, it becomes clear that many new biologically based energy sources will be developed. It becomes equally clear that for the foreseeable future, this will be a diversification of energy sources and not a smooth substitution of one kind of new fuel for old ones. We may well see—at more or less the same time that new, renewable organic sources come on line—an increasing reliance on natural gas.[14]

Brazil took a courageous leap into the biological-fuel era back in the 1970s, but didn't exactly land on its feet. The national government made a decision in 1975—in the depths of the petroleum crisis years when the development prospects looked bleak for any country that had to get its oil on the world market—to replace gasoline with ethanol. Although Brazil lacked known petroleum reserves, it had lots of land and sunshine and people who knew how to grow sugarcane and convert it to alcohol. In less than a decade the country had four hundred industrial plants, half of them linked to sugar refineries, and was producing 9 million tons of alcohol a year. Brazil was saving millions on petroleum imports, and people were calling it the country where the cars ran on rum.[15]

The trouble was, bioethanol was not a clean-burning fuel—at least not in the cars that were using it at the time. The combustion produced great quantities of aldehydes, sweet vapors whose less-than-pleasant smell reeked in the hot, busy streets of the big cities. Its economic performance was also disappointing. The conversion process had been heavily subsidized by the government—according to some estimates, it cost as much as the entire military budget in some years, in an era when Brazil was generous with its generals—and this played a part in the growth of massive deficits. The lowering petroleum prices on the global market didn't help either.

So Brazil and Brazilians paid heavily for that pioneering venture. Call it a learning experience. It was not a complete failure; it did at least reduce the content of lead, carbon monoxide, and other non-sweet-smelling pollutants in the air of the major cities.[16] Over the years the technology improved. Now cars in many other parts of the world are running on biofuels—either ethanol or methanol—produced from various other crops including sweet potatoes and (in the U.S.) corn. A pilot plant in Japan produces alcohol from yeast cells. Other researchers are trying to find better ways to process lignocellulose—the main component of waste products such as wood chips and straw, corn husks and stalks, beet pulp.

This may well turn out to be one of the most promising applications of biotechnology research to economic progress in some of the less-developed countries, since lignocellulose is the most widely distributed, most generally and cheaply available, organic matter on Earth. Better methods of processing biomass could lead to small-scale, village-level factories using cheap material to produce any number of usable chemicals, from fuels to feedstocks for manufacturing.

Another believer in the future of bio-energy is the Japanese government's Ministry of International Trade and Industry. MITI has for some years been funding research to produce hydrogen—the cleanest burning of all fuels, the one many people believe to be the best energy hope for the 21st century—through biotechnology. The present means of making hydrogen is to separate it from water—an expensive and difficult process that burns a great deal of electricity. But numerous microbes and seaweed produce hydrogen naturally, and MITI researchers hope to find or create varieties that can do so in commercial quantities.[17]

Toward the Biological Computer

The computer, the admirable instrument that seems to be capable of rewriting so many of life's rulebooks, is still only a thing made in factories, mostly from familiar industrial-age materials: petroleum-based plastics, metals, silicon. It is a child of the industrial era, even though it has turned out to be a child that—like some offspring of an unfortunate Shakespearean king—tends to take over and change the plot line. Computers have taken over and radically transformed many things, including the computer industry itself and, as we have seen, biology. The industrial revolution led to the computer revolution, which in turn led to the biological revolution. It now begins to appear possible that biotechnology will turn around and transform the computer.

In Japan I went to see Isao Karube, who is known as "the father of the biosensor." He gave me the basic unit from a biosensor—a small device not much larger than a toothpick—that I still keep on a bookshelf in my office, a memento of the past and a clue to the future. It was at about that time—1987—that a few adventurous journalists were just beginning to publish articles about the possibility of a "biochip," a biological computer. I thought the idea a bit dreamy at the time, but Karube took it seriously, believed the work he was then doing might contribute to progress in that direction.

A biosensor combines organic material with an electronic element. When placed in a solution—such as a sample of human blood—the chemical reaction produces an electric signal that can give a precise measurement of a certain substance that might be present. (At this point, they can measure one substance; multisensors will come later.)

About a dozen companies in Japan were already manufacturing biosensors, and they were being used even in such prosaic places as fish markets and breweries. A biosensor can test the freshness of raw fish, or monitor a fermenting batch of beer or sake or soy sauce. In the United States, biosensors are being used to detect narcotics traces in blood or urine samples, and undoubtedly have a promising future in medical applications. It is an established field of bioelectronics, and progress to the biocomputer now seems quite realistic and probable.

Until quite recently the most promising avenue of research led in the direction of using biological molecules—engineered proteins—as the active components in computer circuitry. This would probably not result in a purely biomolecular computer, according to one of the leading researchers in the field, Robert Birge at Syracuse University: "Far more likely, at least for the near future, is the use of hybrid technology in which molecules and semiconductors are used in combination. Such an approach should provide computers that are one fiftieth the size and as much as 100 times faster than current ones."[18]

Computer components made with biodesigned proteins continue to be an attractive prospect, but the proteins were somewhat upstaged late in 1994, when a computer theorist at the University of Southern California, Dr. Leonard Adleman, published an extraordinary article in *Science* in which he (a) proposed a much different approach to the biocomputer challenge, and (b) reported that he had already successfully performed a complex computation—in a test tube. The material he used was nature's own information system: DNA itself.[19]

What he had done was set up a mathematical problem by synthesizing particular molecules of DNA, and then processing the information through the chemical reactions that took place when the molecules were mixed together. The problem was one of those classic mathematical headaches, a search for a path that would connect seven cities with fourteen permitted one-way links between them. It was one of a group of problems that have no simple solutions at all and can usually be solved only by trying out all the possibilities.

Adleman's paper described how the DNA solved the process, and described some of the mind-boggling possibilities of a DNA-based system of computing: able to perform more than a trillion operations per second (a thousand times as fast as the best supercomputer), a billion times more energy-efficient than conventional computers, able to store vast quantities of information. Not surprisingly, his paper touched off an explosion of new interest in biocomputing: a meeting on DNA computers was announced on the Internet, and about 200 excited computer scientists, biologists, and others jammed into a room at Princeton to discuss the new topic.

The consensus at this—very early—stage of DNA computing is that it has astonishing possibilities and, of course, any number of possible hitches and slips. Most likely, instead of replacing conventional computers, DNA may be used to make hybrid computers, or used as information storage units, or perhaps would specialize in certain kinds of huge calculations. But DNA computing, however uncertain its future, seems to have arrived like a lightning bolt. "The floodgates have started to open," one computer scientist said. "I have never seen a field move so fast." Another said: "It's an industry at this point." Another, after naming some of the technical obstacles, said: "If you said to me, 'Here's $10 million. You have one year to build a very fast supercomputer,' I'd probably talk to my friends at Cray." He added: "If you said, 'Here's $10 million dollars,' and gave me five years, I'd probably think of DNA."[20]

Bioconvergences

In the foregoing chapters, we have surveyed biotechnological advances and possibilities in three fields—medicine, agriculture, and industry. We have considered a few specific lines of research—such as gene therapy, tissue culture, biomaterials—that have some promise to be a part of the future. Being neither a scientist nor a fortune-teller, I don't know which will succeed and which will fail. By the time you read this, it's quite likely that some of the projects will have succeeded and some of them will have failed—and equally likely that other surprises (like the DNA computer) will have emerged from somebody's laboratory. Yet, although the specifics are wildly uncertain, the general line of development is not: the revolutions in medicine, the reinvention of agriculture, are happening and will continue to happen.

In relation to industry the picture is less clear. We can expect a slight lessening of the world's obsession with mineral resources and capital, together with a growing respect for the value (and versatility) of genetic resources and the information that makes it possible to use them. These processes may continue at their present modest rate of progress, or one or more of these lines of activity may take a sudden leap forward—technological change has a way of doing that —and surprise us all.

At the present rate of change, the question of how far the greening of industry actually proceeds will depend on how people deal with two kinds of information: economic and ecological. Until fairly recently economic calculations would have completely dominated any dialogue about new approaches to industry. Economic rationality was the common currency, so to speak, of the industrial age, and any question about, say, the future of biomaterials would be sensibly and finally answered by how well they competed in the marketplace. Now we are compelled to think also about their ecological impacts—or about the ecological impacts of the materials, such as plastics, they aspire to replace. The economic criteria certainly have not faded away with an apologetic tip of the hat, but they are now part of a more complex dialogue.

In the matter of bio-energy, for example, some experts are enthusiastic about black cottonwood trees as a future fuel source. They grow fast, and can be converted into methanol, a form of alcohol that works in existing automobile engines. They can be used to "grow fuel"—and perhaps even to grow it at market prices, thereby satisfying the economic test. But the real reason for the enthusiasm about them is that, theoretically, a biomass fuel system based on their use could be nearly "carbon-neutral." That is, the growing trees would subtract from the atmosphere about as much carbon as would be returned when the fuels were burned, and there would be no net increase of greenhouse gases.[21] The calculations are based not only on awareness of the fuel's performance in an economic system, but also of its performance in an ecosystem—the Earth.

That sort of thinking requires several kinds of information that are still fairly new to us: a general image of the world as a biosphere, projections about possible future climate change, sophisticated calculations of symmetrical exchanges of carbon with the atmosphere. These kinds of information—however imperfect they may be, however controversial—increasingly find their way into the public dialogue. We are living in a different kind of world, a global bio-information society, which presents not only technological change but changing understanding of what kind of a world we inhabit and how it works. It is a *cybernating* world, continually giving us feedback about the state of biological systems—and power to influence them.

PART FOUR

THE CHANGING LIFE OF THE PLANET EARTH

CHAPTER TEN

Human Governance of Natural Biosystems

Whether we are considering a simple electric oven, a chain of retail shops monitored by a computer, a sleeping cat, an ecosystem, or Gaia herself, so long as we are considering something which is adaptive, capable of harvesting information and of storing experience and knowledge, then its study is a matter of cybernetics and what is studied can be called a 'system.'

—James Lovelock[1]

We human beings are in fact managing the entire planet Earth, every square centimetre, right now, and the illusion that we are not, that any one of us can be exempt from this task, is extremely dangerous.

—Peter Raven[2]

The Gaia hypothesis offered some years ago by James Lovelock and Lynn Margulis—the idea that the biosphere is a self-regulating entity with natural feedback mechanisms that keep it livable by controlling the chemical and physical environment—has been widely discussed, and in the process has become a part of global culture almost equal in importance to the famed photograph of the planet itself. It has not only been widely discussed, but also widely distorted: distorted in different directions by partisans on opposite sides of the current environmental debates.

People of the not-to-worry persuasion take it to mean that it really doesn't matter how much anybody pollutes, because good old Gaia will clean up after us. Green extremists have turned it into a kind of eco-spiritualism, using the mythic image of Gaia to drive a wedge between the planet and the people who live on it. The former gets to be the wise goddess keeping things nice for the birds and beasts, while human beings are cast in the role of vicious despoilers whose actions only serve to obstruct Gaia in her work. Neither of these versions is particularly faithful to the original hypothesis, and—more to the point here—neither accurately represents the way the world works now. We can't be sanguine about our own messiness, and neither can we leave human action out of the biosphere-regulating equation.

Human beings now play a major role in the management of the world's natural systems—the air, the oceans, the land ecosystems, the wildlife populations, even the global gene pool. Furthermore, such management is not about to cease or even diminish, regardless of how many sermons of hands-off environmental humility are preached at us. In fact, present rates of human population growth and technological progress guarantee that it will increase.

Once upon a time, the world really was a nonhuman system of ecosystems, weather systems, and geological systems, and in that respect it more or less resembled the paradise of the Gaian purists. But the course of evolution has modified those systems in many ways as people migrated, hunted, fished, farmed, irrigated, traded, burned, built, mined, invented. It's a bit of an oversimplification even to say that people have impacts on ecosystems. The statement is quite true, but it does not really give us an adequate picture of how the world works, how evolution has evolved, and how deeply we are involved in Gaia's activities.

Human beings have not only altered Earth's various systems, but have also created entirely new systems. Farms—agro-ecosystems—are one type of system that human beings create. Cities are another. We also have transportation systems, water-delivery systems, governmental systems such as nation-states, social systems such as families and tribes, economic systems such as local and global markets, information systems such as the satellites and the Internet. All these become connected to other systems, human and nonhuman. The vast water system in California connects the town where I live (and my house, and my body, and my plants and pets) to distant snowy mountains, and also connects those mountains to farms in Fresno County and swimming pools in Beverly Hills. As systems become connected, they change; they become different systems, in many ways. One of the most important changes has been the fading of the boundary between the human systems and the nonhuman systems. The only way to have a really clear idea of such a boundary is to remain ignorant of what is going on. And that becomes harder to do, too, because a lot of the systems we create are information systems that tell us what is going on.

This information always has to do with power. It tells us of decisions that have been made in the past, and of the impacts they have had, and it sets us up to make other decisions that have further ecological/biological consequences: water the lawn or let the grass die; vote for the park bill or vote against it; have a baby or take the pill. Every single one of those decisions—from the small choices made by a peasant farmer plowing a field to the global-scale choices made by an international conference on greenhouse gases—is a part of how the world works. Whatever Gaia does—badly or well—she does it with human help. The bio-information society is a global society, and it is actively engaged in global management. And—this is where things really get interesting—it knows it.

I wouldn't say that this knowledge—a general awareness of how human choices have become an integral part of the fabric of the biosphere and its workings and its evolution—has yet reached the stage of great clarity or wisdom. We don't always know what we know. But it becomes increasingly difficult to remain ignorant of the basic fact that human systems have global-scale ecological/biological impacts. Human systems are now part of the governance mechanism of the biosphere as a whole.

When we speak of governance, it is good to remember that governance of large and complex systems is never complete control—not even under totalitarian regimes that have tried mightily to make it that. Human beings play a governance role in relation to nonhuman biological systems, but we remain monumentally ignorant about, and powerless to affect, many details of the biosphere's workings. It is a huge and intricate system whose actions cannot be completely mapped, planned, predicted or directed. We do well to remember that the word *governance*—like the word *cybernetics*—comes from the Greek *kubernetes,* meaning helmsman. Governance is not so much control as the application of feedback that affects the workings of a system. Norbert Wiener, the mathematician who popularized cybernetics in the 1940s, pointed out that information in the form of feedback is at work constantly in the actions of both mechanical and biological systems. Reach for a pencil, he would say, and you are not simply directing a single muscular action but setting in motion a process that will be regulated by feedback—usually in the form of visual information that you use to correct the course of your hand as it performs the operation.[3] He also pointed out that there are all kinds of faulty feedback, communications failures, and other mishaps that can make cybernetic systems overcontrol, undercontrol, or go off in wrong directions. We find that human actions steer the course of evolution now for many species and in a sense for the planet itself, but even small artificial systems—notably the garden plot in my back yard—are not fully under human control.

A tricky situation to be in, this place the human species now occupies. One might reasonably suggest that maybe we ought to put off the governing until we all have a fuller understanding of it. But people everywhere are already governing ecosystems—or to put it another way, ecosystems everywhere are being governed by human systems. There is, for example, no population of wildlife anywhere on the planet that is not subject to human decisionmaking—either directly managed, as in the case of small populations of endangered species, or indirectly affected by the various effluents of human civilization.

Managing Wildlife and Not-So-Wildlife

The words *wildlife* and *management* may sound like they don't belong together at all, but in fact wildlife management is a well-established science and profession—and one that grows increasingly more important. All populations of wildlife are now parts of the global bio-information society. Their movements are studied, their numbers are entered into the data banks, their genomes are mapped and sequenced, and their evolutionary destinies debated by contending groups of scientists, environmentalists, developers, landowners, and public officials. This informatization of wildlife is perhaps most dramatically apparent in the case of the elephant herds being tracked by satellite and watched on monitors in New York City—but in many other ways, wildlife biology is converging with new developments in information and communications technology.

The whole meaning and purpose of wildlife management is undergoing rapid change. In the past it was mostly about keeping populations of animals safe and healthy in order that they could be shot at by sportsmen. Until quite recently the term *wildlife* was interchangeable with *game*. The wildlife preserves were the private forests of kings, nobles, and millionaires, and the job of gamekeepers was mainly to protect animals from unauthorized hunters—i.e., poor people—who wanted to kill them for food rather than for the pure aristocratic fun of it. Much wildlife management is still done by fish and game agencies, and this is often a point of contention now since a new ethic of wildlife management—based on the idea of protecting animals because of their own inherent God-given or Gaia-given value—has won human hearts and minds and is increasingly likely to be a part of the value system of the people who are doing the managing.

Wildlife management in practice has generally included not merely protecting native species, but introducing new ones. In ancient times, kings would import new species of game, and occasionally send breeding pairs abroad as royal gifts. Some animals, such as the pheasant, became familiar far from their original habitat because people liked to hunt them. A curious Darwinian twist; survival of the fittest to be shot at. Introduction of new populations is still a common part of wildlife management—and often a controversial one now, because so many more people become involved in the process.

Operation Moose Lift, a Canadian-American project undertaken in the mid-1980s to establish a herd in Northern Michigan, is one example of such an operation. The moose is not exactly an exotic species in Michigan—there was a time when its natural range extended as far south as the Carolinas—but it has been extinct there for over a century. Some years back a wildlife biologist observed that the forests of the state's Upper Peninsula were recovering nicely from the days when they had been heavily logged over, and proposed the restoration project. Canadian officials offered to provide a startup herd from the Algonquin Provincial Park in Ontario, over six hundred miles northeast of the peninsula.

The actual moose lift involved using a helicopter to herd animals out of the Algonquin forest to an open area, shooting them with tranquilizing darts, wrapping them in large nylon slings and transporting them by helicopter a dozen miles or so to flatbed trucks—then putting them into crates, driving them to Michigan and releasing them into the wild. Many, upon being released, were fitted with radio collars so the movement of the herd in its new habitat could be monitored. The transplantation proved to be a success, and over the subsequent ten-year period the Michigan population increased from the original 59 to over 250.

Operation Moose Lift, relatively noncontroversial as these things go, was financed largely by sportsmen's clubs and enthusiastically supported by local residents, some of whom decorated their cars with bumper stickers boasting of "MICHIGAN—WHERE THE MOOSE RUN LOOSE."[4] It's quite common in wildlife-management cases, however, that different groups become angrily divided about what should be done. One of the most heated of these disputes concerned another Canadian-American project, the restoration of wolves in Yellowstone Park and central Idaho. Environmentalists had championed this idea for years, and were delighted when it finally got under way. "The wolf is the embodiment of wildness, and Yellowstone is the symbol of wild places," said the director of the Wolf Fund in Wyoming. "It's like returning the heartbeat to the heart." But a Wyoming rancher, speaking for a different point of view and also expressing the deep anti-federal-government sentiment that is rampant in those parts, said: "The issue is not wolves. The issue is control of the land. This is part of a bigger agenda from the Interior Department to control the West. If they control the land and if they control the water, then they control the people."[5] Fearing that the wolves would decimate livestock, opposition groups filed a lawsuit, and the future of the wolf pack was placed in the hands of a federal district judge. It is frequently jurists and bureaucrats, not the abstract rules of survival of the fittest, who actually determine the evolutionary fortunes of wild animals.

Another issue in wildlife management—somewhat more subtle than the primeval wolves-vs.-cattle calculation—is the question of just how wild wildlife really is when it lives within the protection of human beings. Some critics of the long and painstaking effort to restore populations of the giant California condors say the condor-hatching program is also a sort of behavior-modification program. Condors range over great distances and feed on carrion; frequently they find and eat carcasses of animals that have been shot by hunters, and are full of toxic lead shot—which was one of the main reasons the condors became endangered to begin with. Wildlife biologists proposed to put out lead-free meat for the condors in selected feeding areas, and to train them to return to those areas. This was strongly criticized by some environmentalists because it appeared to be a case of "tinkering with nature."[6] The alternative to tinkering is to manage the habitat in such a way that the condors might live as if there were no human beings around—which is hard to bring off when the habitat is a range of mountains in the most populated state in the country, with freeways to the east and west of it.

Let us look at one more example, in which it becomes particularly clear that wildlife management is also evolutionary policy: the case of the Florida panther. The panthers roam in the wilds of northern Florida, in and around the Osceola National Forest and across the border into Georgia's Okefenokee Swamp. They are protected under the Endangered Species Act and their numbers are watched closely. In 1993 the Captive Breeding Specialist Group, a private scientific think tank, studied the data collected by the Florida Panther Interagency Committee (a management group representing four federal and state agencies) and issued a report warning that unless something were done, reproductive problems and other genetic defects—such as heart murmurs—would be made worse by continued inbreeding. The panthers, then numbering about 50 adults, would become extinct in less than 25 years. Note, here, that the critical information was in the form of a projection based on genetic analysis. The scientists were asking the officials to respond to something that hadn't happened yet.

So the matter was bucked upstairs to the appropriate evolutionary-governance agency—in this case the Division of Endangered Species at the U.S. Department of Interior in Washington, D.C.—where the officials decided to bring in some fresh breeding stock. Texas cougars are a close relative of Florida panthers, and ten of them—some captured in the wild, some bred in captivity—were brought to Florida, fixed with radio collars so that their movements could be tracked from airplanes flying over the area, and released. The director of the agency acknowledged that this was a fairly drastic measure, since it would probably "reconstitute the original genetic diversity" of the panther population. But it had worked with other species and—a most important point here—it would not jeopardize the panthers' legal status: Texas cougars and Florida panthers were considered subspecies of the same species, had interbred in the not-too-distant past, and the panthers would still be officially a distinct (and endangered) species, not a hybrid.[7]

In late spring of 1955, shortly after the first Texas cougars had been released, the signals emanating from the hot Florida forest showed that male Florida panther No. 54 and female Texas cougar TX-106 were spending quite a lot of time together and that the beeps from TX-106 would most likely soon indicate that she had stopped roaming and begun to prepare a den for kittens. The Florida wildlife biologist overseeing the restoration program was most enthusiastic—not just about the immediate results, but about the potential usefulness of this particular bio-information convergence. "This project probably wouldn't even have been conceived of ten years ago," he said. "Now we can combine the field aspects that we learn from telemetry with detailed molecular genetics and their medical history. We have an opportunity to really learn something significant, not just about Florida panthers, but about natural biological processes and how we can sustain species in the future."[8]

In one way and another, human governance and management—either active management as in the case above, or passive management by protecting species and their habitats—continue to increase. "Passive" management produces its own kind of actions, such as patrolling the boundaries around parks and wilderness areas and enforcing the regulations within them. And it is not only a matter of dealing with pieces of real estate. People also manage the birds in the air—as when governments negotiate international treaties to protect their migration routes—and the fish in the sea. There is a marine wildlife sanctuary off the coast of California, and on the Atlantic coast, fishermen and conservationists and scientists and government officials are wrangling over the extremely serious depletion of the offshore fisheries. One of the proposed solutions—another form of governance—is to "privatize" the fisheries by assigning individual fishermen the right to catch a fixed percentage of the annual harvest. The idea of this approach, which is already being used with some species of fish, is that the fishermen with quotas have a strong interest in maintaining the health of the fishery as a whole—the more fish, the larger the quotas and the more positive feedback in the form of hard cash. That's the good news. The bad news is that assigning quotas among people whose livelihood depends on getting them is a terribly difficult thing to do, and a lot of fishermen end up out of work. The fate of the underwater ecosystems is inextricably interwoven with the human economic and political systems that govern them.[9]

Biotech in the Wild

Biotechnology has already begun to play a part in wildlife management—indeed, several parts. Among the first medical products based on recombinant DNA technology were animal vaccines. Originally these were intended for domestic animals, but some have proved effective with wild animals as well. France and Belgium fought a rabies epidemic among foxes by air-dropping food laced with a vaccine. This was so successful that the World Health Organization picked up the idea and ran with it—soon air-dropping oral vaccines to wild dogs in Tunisia, Turkey, and parts of southern Africa. Public health experts expressed great enthusiasm about such programs. Some thought that, although there is not much chance that rabies can be wiped out completely, it may well be possible to eliminate it from entire nations or continental regions.[10] Australia has experimented with a genetically engineered virus designed to spread infertility among its notorious rabbit population—another step in the direction of large-scale ecosystem management.

The Bigger Picture: Managing Ecosystems, Protecting Continents

Increasingly the emphasis has shifted to the management of ecosystems rather than the management of populations, and to the protection of endangered ecosystems rather than the protection of individual endangered species within them. There really isn't a clear distinction to be made between ecosystem management (or protection) and wildlife management (or protection); in fact there is a general continuity now from gene banking through zookeeping to wildlife and ecosystem management. We may compartmentalize them in our minds and our bureaucracies, but they are all merely facets of the ever-expanding role that human beings play in the life of the biosphere. Environmentalists know about the importance of ecosystems, and have often used a legally protected endangered species—the spotted owl, the snail darter—as a means to preserve an ecosystem that did not enjoy any such status.

There really isn't any such thing as a truly pristine, hands-off ecosystem in the world, although some wildnerness areas resemble that and are managed so as to avoid obviously destructive impacts such as development, mining, and logging. Preservation, however, is always politics, and always difficult. When I visited Costa Rica some years back, I was impressed by its national park system—which is widely admired by international environmentalists—but was disturbed by the news that the government had to patrol park boundaries and throw out the poor people who invaded to scratch out patches where they could plant corn on the hillsides. In South Africa democratic reforms are threatening the great 7,000-square-mile Kruger National Park, which is surrounded by settlements of the country's poorest people—some of whom would like to graze their animals there, others of whom demand at least a role in its management.[11] Sometimes these dilemmas can be resolved: Costa Rica, for example, has made progress in "agroforestry" projects that enable local people to extract a living from ecosystems while maintaining them in relatively pristine condition. But the development-preservation tension is an ever-present troublemaker in ecosystem management.

There are also, as in the case of wildlife management, controversies about what is the most natural way to manage nature. Probably the hottest issue—no pun intended—is the matter of forest fires. The huge Yellowstone fire, actually a series of fires, in the drought summer of 1988, brought to public attention a huge controversy among forest-management people about whether fires should be allowed to burn or be put out. Allowing them to burn is generally considered the more natural response, but it conflicts with the views of homeowners in or near the forests, people who think national parks should be protected, and officials who understandably want to get in there and *do* something when a forest is going up in flames. Advocates of "natural burning" say fires should not only be allowed to happen, but should be *made* to happen from time to time to prevent buildup of debris that will make a fire more dangerous when it occurs accidentally.[12]

The many technical and political difficulties are complicated by the fact that any given ecosystem is in turn part of still larger systems. Wetlands, whose protection is being debated in the U.S. Congress at the time of this writing, serve not only as habitat for the species within them, but also as necessary way-stations for migrating birds. They also perform functions for the larger river systems and watersheds to which they are connected, by filtering impurities from water and controlling floods. One of the most important wetlands regions in the United States is the Everglades region of South Florida, which is now slated for a massive replumbing effort, likely to take fifteen or twenty years and to cost some $2 billion. Restoring wildlife habitat is one of the many goals of this project, but it also has to do with replenishing the underground aquifers that are the water-supply source for the huge—and growing—urban regions along the coast. The project has generated its share of political conflicts, but it's generally understood as one that has payoffs both for environmental protection and urban development. Considering the size of the area involved—some 14,000 square kilometers—and its proximity to big cities and rich farmlands, this is one of the most important habitat-management plans that has ever been devised.

Despite the popularity of the "small is beautiful" slogan, large is usually more beautiful when it comes to managing and protecting more-or-less natural areas. Smaller parcels tend to be more fragile, more easily damaged by activities in surrounding areas. This was the logic that led to the establishment of the world's largest wilderness area, the 25 million-acre Tatshenshine-Alsek Provincial Park in northern British Columbia. It could be argued that the largest wilderness area of all is the continent of Antarctica, which was given an internationally protected status in 1959. Under the treaty various nations that had claims on some part of the continent agreed to suspend their claims and open it up for scientific research—and to proscribe any military activity, nuclear tests, or disposal of nuclear wastes.

Actually, all of the world is subject to human management in one way or another. Some of the management, however, is less direct—such as the human impacts on the atmosphere. That is the larger and somewhat frightening truth that is being revealed by the debate about global warming.

An Even Bigger Picture: Governing Gaia

It was just about at midpoint in this century that, thanks to the atomic bomb, people first began to get acquainted with the idea that something catastrophic could happen to the whole world as a result of human action. Compared to that prospect, all previous wars seemed relatively puny events. Most of us who are alive now have little memory of a time before that frightening possibility emerged—before those deep images of a single biosphere, a vulnerable world, a world that could be permanently altered by something people did, became a part of our common consciousness. We lived with that for decades, and then the prospect of nuclear holocaust began to lose some of its fearful power at just about the time that the possibility of a different kind of global holocaust —new images of a single biosphere, a vulnerable world, a world capable of being permanently altered by human action—began to emerge. It has not been a really comfortable fifty years.

The scenario of global warming entered the public consciousness quickly and vividly. Actually, it had been back around the turn of the century that the Swedish scientist Svante Arrhenius first predicted that increases of carbon dioxide in the atmosphere would lead to an increase of as much as nine degrees in the average global temperature. But it was a good deal later than that—in the late 1980s, when a heat wave hit the American Midwest and East Coast—that the once-obscure speculation revived with a vengeance, became a subject of Senate hearings and front-page newspaper articles. The scenarios that went with it were ghastly predictions of blistering heat, polar ice caps melting, oceans rising, weather systems going berserk. At about the same time, environmentalists were drawing public attention to depletion of the ozone layer caused by human use of chlorofluorocarbons. The environmental organization Greenpeace ran full-page newspaper ads warning: "Normal life could be interrupted for generations. In some cases it could be dangerous ever to go outside." The new disaster scenarios were endorsed by important and serious people. No less a public figure than George J. Mitchell, then Senate majority leader, wrote:

> Climate extremes would trigger meteorological chaos—raging hurricanes such as we have never seen, capable of killing millions of people; uncommonly long, record-breaking heat waves; and profound drought that could drive Africa and the entire Indian subcontinent over the edge into mass starvation. . . . Even if we could stop all greenhouse gas emissions today, we would still be committed to a temperature increase worldwide of two to four degrees Fahrenheit by the middle of the twenty-first century. It would be warmer then than it has been for the past two million years. Unchecked it would match nuclear war in its potential for devastation.[13]

Then, inevitably, the backlash began. Conservative writers sprang to their word processors and hastened to assure us that global warming was not going to happen, or that, even if it did happen, we would learn to love it. A publication of the Hoover Institution, the right-leaning think tank at Stanford, called global warming "a boon to humans and other animals," and pointed out a number of beneficial effects to be considered alongside the problems. A warm climate would lower transportation costs, because snow and ice hamper the movement of cars and trucks and storms disrupt air travel. New tourist opportunities might develop in places such as Alaska and northern Canada. There would be fewer power outages and interruption of wired communications. The southwestern United States would become wetter and hence better for agriculture.[14]

The bias in the various reactions to the climate change issue is all too obvious, and hardly encouraging for the future of the global bio-information society. Human beings, it appears, filter their feedback through their ideologies. The view from the right has always been that the data is unpersuasive, the predictions of global warming inaccurate —or that, if they are accurate, it's no problem anyway. No cause for tinkering with the economy. The view from the left is that the disaster is either coming or already here, and that the world should immediately do any of a number of things that people on the left have been saying it should do anyway, such as regulate industry more strictly and switch to alternative energy sources. At a conference on global warming held by the American Association for the Advancement of Science I heard a panelist say that there had been a lot of solutions running around in search of a problem, and that they seemed to have found it. At about that same time I read an article by a Marxist in which the writer expressed his hope that the climate change issue would "save socialism." More recently, a writer in *Nature* said: "In Britain . . . separate groups pursuing commercial interests, foreign policy goals and domestic politics each discovered their own uses for the warming hypothesis. For some, the opportunities included the pursuit of global scientific research agendas, for others, the enhancement of bureaucratic power at home. . . . Calls for environmental regulation were generally attractive to environmental bureaucracies . . . beleaguered national politicians gained a world stage on which to indulge in global green rhetoric without, as yet, having to face issues of domestic implementation."[15]

Yet, for all the conflicting data about the climate change and ozone depletion questions—and even more conflicting opinion about how to interpret it—a larger truth is becoming visible as people become accustomed to a more systemic view of the world. The ideas of a connection between chlorofluorocarbon uses and the ozone layer, between the production of greenhouse gases and the world's climate—these are widely held now, and they form a fundamentally new and different way for human beings to understand the biosphere. And when we read in the papers about international conferences and agreements—or failures to agree—on widespread policy changes, we absorb an even more revolutionary concept: the idea of governments taking action to protect the atmosphere and the global climate. Even the conservative dissent contributes to this to some extent, by debating the necessity of such action. Sometimes what you are arguing about is as important as who wins the argument.

We have come to the place—rather speedily, I would say, when you consider how recently our ancestors wandered naked on the Olduvai Plain—where human activity has changed the world's atmosphere. This much—CO_2 buildup—is rather well documented, as is the likelihood of its increase. The results are still much in doubt, but the scientific consensus is moving toward the expectation of global warming and rises in the sealevel. We have also reached the point where information systems tell us more and more about these developments as they unfold, and where our communications systems carry the news around the world, and where people everywhere debate possible governmental response. We have before us an issue of ecological management on a global scale.

Paul Wapner, writing in the journal *Politics and the Life Sciences,* makes a useful distinction between global problems and world problems. He defines a global problem as "one that physically affects everyone throughout the world," and offers the greenhouse effect, ozone depletion, and global species extinction as three examples. He says that, in contrast, world problems such as security, human rights, and hunger "do not affect everyone on earth. In fact, many people care little about them or are simply unaware of them."[16] The emergence of truly global problems indicates, he says, "that human experience in the late twentieth century is distinctly different from that of previous times. They signify humanity's entrance into the global age."[17]

Personally I would prefer to call them issues rather than problems, because, in the case of each of Wapner's examples—greenhouse effect, ozone depletion, global species loss—people don't agree unanimously that there is a problem: that's the issue. But they are definitely *global* issues, and thus preludes to the politics of the twenty-first century—the global politics of a world increasingly wired and continually wary about signs of disturbance in the biospheric systems that sustain all our lives.

It is easy to become discouraged by the posturing and hyperbole on both sides of these issues, but on the other hand it is worth recognizing that we have come a long way: we have information systems that can measure parts per billion of ozone, and governance systems that can respond with international commitments to abolish the manufacture of CFCs. In the more complex case of the greenhouse effect, we have information systems that measure the buildup of CO_2 and other gases in the atmosphere, climate models to test hypothethes of the possible effects, and international gatherings such as the Montreal conference on CFC reduction, the Earth Summit in Rio and the Berlin conference on global reductions of greenhouse gas emissions. Nature lovers no doubt would prefer that such matters be determined by the serene wisdom of Gaia's nonpolitical systems, and others of a more technical bent might prefer that the scientists and their instruments make the decisions and the rest of us shut up about it—but, for better or for worse, that is not the way it works now. Human systems and the workings of the biosphere cannot be torn apart, and major ecosystem-managing decisions—especially the big ones—are in the realm of politics.

Depend on the global information systems to keep biosphere management on the public agenda. The U.S. government's Global Change Research Program currently orchestrates the work of eleven different federal agencies; among those on the team are the National Science Foundation, the National Aeronautics and Space Administration, and the National Oceanic and Atmospheric Administration. The work of the researchers is augmented by satellites in space, observation stations on the ground and on the oceans, various air and water sampling devices, countless gigabytes of data, and state-of-the art computers and communications networks. *Earthwatch* magazine calls this "unquestionably the most ambitious and comprehensive Earth-monitoring program on the planet," but adds that "it is only part of an international phalanx of similar efforts from dozens of other countries, including an American Total Ozone Mapping Spectrometer aboard a Russian Meteor 3 spacecraft, a U.S.-French Topex-Poseidon satellite to measure global sea level rise with an accuracy of two centimeters, a German-built Shuttle Pallet Satellite (to measure atmospheric infrared and the critical chemical radical OH), a Japanese-American test of the NASA Scatterometer, and dozens of other esoteric and complex measuring devices involving dozens of countries and an absolute forest of acronyms." The same report notes that the biggest part of the whole project, scheduled to come on line in 1998, is NASA's Earth Observation System (EOS), with a series of satellites launched over a ten-year period to measure everything from aerosols to phytoplankton.[18] Whatever happens will not go unnoticed.

Meanwhile, there is talk of yet another global problem. A group of about 150 astronomers and space researchers gathered in California in 1995 to consider whether an asteroid or comet might hit the Earth. The general conclusion was that the possibility is remote, but not so remote that it can safely be ignored. It has probably happened on Earth before, and it definitely happened not long ago on Jupiter when the Shoemaker-Levy comet broke apart and plunged into the planet's atmosphere with the force of several million thermonuclear bombs.[19] Information from two different scientific explorations—one reaching into the Earth's past, the other observing events in the outer reaches of the solar system—have given us something else to worry about. Speakers at the conference discussed the idea of converting intercontinental ballistic missiles to antiasteroid weapons, and the group's report to Congress recommended action now to intensify detection systems using telescopes and radar: a little more wiring of the world, yet another reinforcement of the image of a single biosphere, a vulnerable world, a world capable of being permanently altered—or in this case protected—by human action.

What Kind of a System?

As we look at some of these examples of what is going on in the biosphere today, it becomes quite obvious that the world's ecosystems are no longer operating separately from human agency. As people try to understand this world, they naturally turn to various concepts of systems—the general systems theories of people like Ludwig von Bertalanffy and Kenneth Boulding, the cybernetics of Norbert Wiener and his colleagues, the Gaia hypothesis and the more recent systems thinking of the complexity theorists at the Santa Fe Institute.

But what kind of a system is this planet with all its nonhuman life forms, its billions of human beings, its global bio-information society and its electronic noosphere of Earth-watching technologies? It is, certainly, a *natural* system, a living biosphere that had a long, rich existence before it began to people itself. But it is also now in a sense an *artificial* system with its managed ecosystems and electronic feedback loops. And, at the same time that we grow increasingly aware of its *unity*—its life as a single system—we become aware of its astonishing *diversity* of subsystems and subsystems within the subsystems. And it becomes equally obvious that, although the human role in the working of this system can reasonably be called "management" or "governance," it isn't a simple mechanistic kind of governance in which everything is neatly planned, organized and directed. This may disturb you—unless you consider the evidence that systems of all kinds operate by organizational principles that don't have much in common with those neat rectangle-and-straight-line diagrams bureaucrats go in for.

The complexity theorists are especially interested in the workings of what they call "complex adaptive systems"—the kind that can acquire information about the environment and their actions within the environment, use that information to create a sort of map, and then act on the basis of it.[20] Not all systems are of this type. A thermostat can respond to feedback from its environment, but it doesn't learn or change its behavior. Neither, really, does a more elegant and complex "expert system," which has information pumped into it by human experts and can perform impressive tasks such as diagnosing illnesses. As physicist and Santa Fe Institute cofounder Murray Gell-Mann points out, such a system "does not learn more and more about diagnosis from its experience with successive patients. It continues to use the same internal model developed by consulting the experts."[21] But true complex adaptive systems are at work around and within us. Some biological examples are people, brains, immune systems, ecosystems, and cells. Some cultural/social examples are political parties, nations, scientific communities. The more advanced robots and "artificial life" computer programs also qualify. All these systems, according to the computer scientist/ evolutionary theorist John Holland, have certain features in common:[22]

First, each system is a network of many "agents," more or less doing their own things: the cells in a brain, the species in an ecosystem. Control tends to be highly dispersed throughout such a network. This is true even of organizations that are supposed to be hierarchical (just ask the President of the United States), and is also true of a complex system such as your own body. We could not live if we had to rely on our conscious brains to trigger each breath and heartbeat, monitor and manage all the actions of organs and glands.

Second, a complex adaptive system has many levels of organization, with agents at any one level serving as the building blocks for agents at a higher level. A group of proteins, lipids, and other small citizens make a cell, a group of cells make a tissue, a bunch of tissues make an organ. Similar sets of systems and subsystems can be found in factories, or in economies. Also—and most important—complex adaptive systems are always changing things around, revising and rearranging all these pieces in response to information, in search of more efficient ways to operate. Although the examples of such systems I have given are nouns, I think it is really closer to the spirit of complexity theory to think of them as verbs—not things, but things happening. Patterns of interaction, flows of information.

Third, all complex adaptive systems anticipate the future. You might not think of a bacterium as a futurist, but every living creature has within its genes an implicit prediction of the future. Farther up the evolutionary ladder, every creature with a brain has many predictions based on learning from prior experience: you know that if you put your hand in the fire, it will hurt. These predictions create behavior in the system, and are continually being revised in response to new experience.

Fourth, complex adaptive systems have many niches—occupations, if you will, that agents within the system are always adapting themselves to fill. And whenever an agent fills a niche—whenever some entrepreneur starts a new business, or an animal finds a different strategy for eating and reproducing—still more niches are created for new parasites, predators, partners. The system as a whole is always evolving, creating new job opportunities. And living things within it are always looking for work. You would naturally expect, then, that many people and groups would colonize new issues such as global climate change, and that some lawyers would discover great possibilities in new products such as Norplant.

What does this tell us about human governance in the biosphere? Quite a lot, I think. It doesn't really grapple with politics, conflict, and power—subjects I will say more about in the next two chapters—but it does remind us that the human decisions on ecological management issues, even global ones, will be made in many places.

It is not surprising that so many people who try to wrap their minds about this emerging global situation are drawn toward theorists like Ilya Prigogine, the brilliant Belgian physicist whose exploration of "spontaneous self-organization" in all kinds of systems—living and nonliving—suggests that people can cooperate without having orders handed down to them through some global chain of command.[23] Such ideas, when taken out of their scientific context, sometimes sound like a prescription to lean back in New Age bliss and stop worrying about what the government does—but I think there is much in them that deserves serious attention. They are fragments of a new worldview that is trying to take form as people recognize a new world.

Lacking in this worldview—so far—is a recognition of the complete inseparability of natural systems from human systems, and of the ways human governance of Earth's systems is being transformed. The idea that we are a part of the Earth—not just on it—is a staple of many ecological and spiritual traditions. The idea that we are inseparable from our inventions is being advanced by a number of thinkers, as we saw in Chapter 4. But these ideas are not yet taken deeply into our consciousness. We don't yet have a way of talking about this planet and its people, organizations, communications systems—this planet in the process of changing fundamentally, reinventing itself—as the single entity that it surely is. We lack a science of such a complex adaptive system. We don't really even have a good evocative metaphor for such a bionic, evolving Gaia.

CHAPTER ELEVEN

Changing Shades of Green

If you love environmentalists, as you should, today the greatest favor you can do to them is to toss cold water on their heads.

—Gregg Easterbrook[1]

Nature in the twenty-first century will be a nature that we make; the question is the degree to which this molding will be intentional or unintentional, desirable or undesirable.

—Daniel Botkin[2]

Many of the matters we have considered in the past chapter come under the general heading of environmental issues, although I am not at all sure how far environmentalism will take us in the future. Its weaknesses as ideology and philosophy are becoming glaringly apparent now—and yet the historical importance of the modern environmental movement can scarcely be overstated. Its emergence in this century was a landmark event in human evolution, an indication that *Homo sapiens* was discovering the world and discovering itself in a new way.

Species-wide discoveries don't happen all at once, however, although the time required for such information shifts is becoming much, much shorter. It took quite a long while, after the mapmakers and navigators had discovered the spheroid world, for the image of that new world to settle fully into the consciousness of all the people on it. Actually, I don't think many people, even in the more scientifically advanced countries, really absorbed the idea of Earth as a planet until the astronauts took a picture of it. It was, perhaps coincidentally, at about the time of the Apollo mission that people in large numbers began to discover the extent of human impacts on the world—the environmental discovery.

That discovery had actually begun in the late nineteenth century. An American scholar, George Perkins Marsh, was the godfather of contemporary environmental awareness. In his monumental 1864 book *Man and Nature, or, Physical Geography as Modified by Human Action,* he summarized most of what was then known about how the world is transformed by ordinary human activities such as farming, logging, firebuilding, modifying waterways, breeding, and moving domestic animals.[3] Nobody before Marsh had ever put those pieces together, shown so powerfully how human beings alter (and damage) the world around them, and his book was enormously influential. He was the main intellectual inspiration behind the early conservation movement —the John Muir and Teddy Roosevelt movement, the forerunner of contemporary environmentalism—that flourished in the United States and Europe in the early 1900s. He did as much to map the world we live in as did Darwin or Mendel or Watson and Crick.

Marsh's message, somewhat like that of Arrhenius in Sweden, receded from view in the mid-twentieth century, and then began to be revived in recent decades. It's hard to give precise reasons for the ups and downs of causes and ideas, but surely changing information technology had a lot to do with it. More data about environmental impacts was being collected, and the word was getting around. I date my own enrollment in the movement from the summer of 1964, when I heard Senator Gaylord Nelson of Wisconsin give a rousing speech about air and water pollution. The senator's rhetoric had exactly the impact on me that he had meant it to have. I went home and joined the Sierra Club—one of the relatively few environmental organizations in existence at the time—and started writing articles, organizing meetings, and leading wilderness trips. A few years later I produced a book entitled *Politics and Environment,* one of the first college texts on environmental issues.[4] At about that time, amid a blaze of global media publicity, came the first Earth Day, and a new surge of environmental activism that has continued until the present time. So I have watched the environmental movement from inside it for thirty-some years, and it becomes apparent to me now that, although the movement is still strong, it is going through a major transition, changing and dividing like a complex adaptive system—and that it needs to go much, much further still.

In the 1970s and 1980s I saw environmentalism become not only a compendium of worries about resources and pollution and open space, but something more—a rejection of the idea of progress, of the whole course of Western civilization. A new troupe of green ideologues emerged, telling us that our whole history had been a series of mistakes, a headlong charge in the wrong direction that needed to be reversed by a mighty, revolutionary new act of human will. Thomas Berry, in a fairly typical—and relatively moderate—expression of this view, declared:

> If the industrial economy (which has well nigh done us in) in its full effects has been such a massive revolutionary experience for the earth and the entire living community, then the termination of this industrial devastation and the inauguration of a more sustainable lifestyle must be of a proportionate order of magnitude.
>
> The industrial age itself, as we have known it, can be described as a period of technological entrancement, an altered state of consciousness, a mental fixation that alone can explain how we came to ruin our air and water and soil and to severely damage all our basic life systems under the illusion that this was "progress."
>
> But now that the trance is passing, we have before us the task of structuring a human mode of life within the complex of the biological communities of the earth.[5]

This statement contains two of the main themes of environmental thought: the critique of industrial society, and the dream of a radically different human civilization able to fit itself quietly and unobtrusively within the prehuman global system of ecosystems. Both are valuable concepts, but both have proved to be controversial and divisive.

The first part, the critique of the industrial age, has led to some productive rethinking and reengineering, but has divided environmentalists—who are not in full agreement that industrial civilization is pathological and incurable. It has also produced antienvironmentalist critiques and backlash.

The second part, the quest for a way to "live lightly on the planet," as the popular slogan goes, has led one wing of the environmental movement into a never-never land of muddled philosophizing from which it may never return.

Clearly industrial societies have been, whatever their positive achievements, hideously destructive to ecosystems—the effluents poured into water, the soot churned out of factory smokestacks, the careless destruction of habitat, the prodigious waste of energy and resources. The big questions are whether such societies are adaptive systems that can learn and change—and clean up after themselves—and whether the course of technological innovation and economic growth doesn't tend toward developing better and cleaner forms of industry and agriculture anyway. Environmentalists are deeply and often heatedly divided about such matters, so deeply that environmentalism is not really a single movement any more—if, indeed, it ever was. The designation is applied equally and carelessly to big-time professional lobbyists and to grassroots "not in my backyard" activists—groups that are quite different, and frequently at odds with one another. It is applied to optimists and pessimists, radicals and conservatives, big-game hunters and animal-rights defenders. It is also applied to movie stars, intellectuals, nature lovers, anarchists, misanthropes, ecosaboteurs, ecofeminists, ecopsychologists and ecospiritualists, to name but a few more shades of green. One should keep this in mind and avoid thinking of this huge outpouring of thought and action as though it were a single monolithic movement that is headed in a single direction. Criticisms of one flavor of environmentalism don't always apply fairly to all.

Criticisms there have been, and they seem to be coming in increasing strength and numbers. To be sure, opposition to the environmental movement is hardly new. From the time of its global reappearance around 1970 plenty of people—among them political conservatives, industrialists, developers, labor leaders, landowners, and farmers—either disagreed with or detested the new activists with their talk of slower economic growth, new regulations, and centralized land-use planning. But more recently there have been new and different critiques, such as Gregg Easterbrook's 1995 book *A Moment on the Earth,* from people who are essentially friendly to the goals and values of the movement. Yes, I know plenty of environmentalists who think that with friends like Easterbrook you really don't need enemies—but there is a reason his book came along when it did, and was taken seriously. The environmental movement had offered a critique of industrial civilization; the time came for a critique of the critique.

Easterbrook's case was that advanced industrial societies were proving themselves quite capable of responding to environmental problems and, as the saying goes, cleaning up their act. Here, quoted, are a few of his main arguments:

> That the environments of Western countries have been growing cleaner during the very period the public has come to believe they are growing ever more polluted.
>
> That First World industrial countries, considered the scourge of the global environment, are by most measures much cleaner than developing nations.
>
> That most feared environmental catastrophes, such as runaway global warming, are almost certain to be avoided.

> That nearly all technical trends are toward new devices and modes of production that are more efficient, use fewer resources, produce less waste, and cause less ecological disruption than technology of the past.[6]

Easterbrook proposed a new shade of environmentalism, which he called ecorealism, and closed his hefty book with an "ecorealist manifesto." One of its propositions was that the environmental movement can only become stronger by "graduating from overstatement" and developing a better capacity for self-criticism. It will be interesting to see if that happens, if the movement—or at least some significant segments—can enter into a more mature engagement with the bio-information society, and perhaps even take a role of leadership in it.[7] If it does, it will be over the dead bodies of the Far Green—the bioregionalists, deep ecologists, and other voices of nature who have been influential and visible in defining the philosophy of environmentalism.

The Downside of the Good Side

The Far Green isn't exactly a single movement either—there are many variations and factions, and much conflict within it—but despite this internal contentiousness it has a fair degree of coherence. It has taken the form of a political movement in the various Green parties, a spiritual or theological movement both outside and inside established religions ("God is green," proclaimed the Archbishop of Canterbury a few years ago), and a philosophical movement in Deep Ecology and other schools of thought. I think it can most accurately be described as a subculture, with its own stores, magazines, countless books, clothing styles, conferences on the Internet, organizations, and shades of ideology. The ideology is mostly what you might call the strong form of the critique of the industrial era, escalated to an indictment of all human civilization and progress as one sustained, murderous attack upon benign nature. "The most violent and widespread war in human history has been going on for more than 5,000 years," declares an Earth First! activist: "This war is, of course, against the Earth."[8] The Far Green is strongly committed to the agenda of finding a way to inhabit the Earth without having significant impacts upon it. Its theorists tend to advocate a global retreat into an urban intellectual's fantasy of the simple life. Let me mention briefly a few of the key concepts.

Bioregionalism. The basic premise of bioregionalism is that the world is naturally divided into bioregions, and that human beings should live within these—preferably in the country or in small cities, practicing agriculture, minimizing industry and trade, not traveling outside the bioregion any more than necessary, becoming acquainted with its local flora and fauna. This is an appealing vision, one that has caught the hearts of romantics at many times and places, but its shortcomings become apparent when you look at statements of its principles such as the 1985 bioregionalist manifesto *Dwellers in the Land,* by Kirkpatrick Sale. One of its weaknesses is the uncritical acceptance of the idea that if people just stayed in their bioregions, ecological wisdom would flow naturally from them as a result and all would be well:

> People do not, other things being equal, pollute and damage those natural systems on which they depend for life and livelihood if they see directly what is happening, nor voluntarily use up a resource under their feet and before their eyes if they perceive that it is precious, needed, vital; nor kill off species they can see are important for the smooth functioning of the ecosystem. When they look with Gaean eyes and feel a Gaean consciousness, as they can at the bioregional scale, there is no longer any need to worry about the abstruse effluvia of "ethical responses" to the world around.[9]

This belief—that life close to the land leads inevitably to ecological wisdom—is a cherished one among the urban intellectuals who pay most of the dues to environmental organizations, even though archaeologists and historians have provided mountains of evidence that people have often managed to mess up the very ecosystems they had inhabited over a long period of time. Then there's also the embarrassing matter of the tendency of rural people to hate environmentalists. Personally I see nothing wrong at all with staying in one place and getting to know it well, but a good case can be made that what we really need now is a more cosmopolitan ecological consciousness with a highly-developed sense of how things work in other regions, and in the world as a whole.

The other weakness has to do with the question of how the transition to a bioregional society might be achieved. Sale's answer is that, first, you break up all big cities and create a lot of smaller cities, each with a surrounding greenbelt. Then you move everybody either to one of these smaller cities or into the country. "Population relocation of this sort need not be all that difficult or disruptive, stretched over several decades and done with due regard for natural systems."[10] The faults in this vision are many, but I think the most important is its failure to engage the question of power—that is, what you do about the wrong-thinking citizens who don't want to be shipped off to the bioregions that they came to the cities to get away from. This problem with power is a common flaw in Far Green thinking, as becomes even more apparent in the writings of Deep Ecology.

Deep Ecology is the creation of Arne Naess, a Norwegian philosopher whom I met one time in Berkeley, some distance from his bioregion, when he was traveling about expounding his ideas. He told me that he thought the world ought to have no more than 100 million people in it, at the absolute maximum, in order for other living things to flourish again. I have to admit that I found this a not entirely unattractive agenda, as long as I got to be one of those still present, but I wondered how he proposed to carry it out. He was not really prepared to discuss that question—or even, so far as I could tell, particularly interested in it. Later I noted that in the Deep Ecology literature, family-planning is dismissed as "shallow ecology."

Deep Ecology has since been taken up by a pair of California college professors, George Sessions and Bill Devall, who have written and lectured on the subject and contributed greatly to its advancement as a main theme of Far Green thought. I will probably not do justice to it in summarizing its key points, since I think it is neither ecology nor deep, but let me try. First of all comes the rejection of anthropocentrism, the human-centered view of the world. The world must not be dominated by the human species, by human needs and human agendas, but must be a place in which the human species lives humbly among all the other species: no different, certainly no more powerful. What follows from this, logically, is a "not do" philosophy in relation to all environmental issues. Sessions and Devall give an example concerning the policy of allowing ranchers to graze animals on federal land: "A simple 'not do' solution," they wrote, "is to end federal subsidies and restrict use of public lands for grazing to a level consistent with recovery of grasslands as estimated by professional ecologists."[11]

This sounds eminently reasonable, but we should note that two exercises of power are involved here. First, the restriction of use. Somebody has to enforce such a policy, since the people who live in the bioregions in question are not likely to comply voluntarily; in fact, the Gaean consciousness of cattle ranchers being what it is, somebody had better be prepared to enforce it at gunpoint. Second, a bit less obvious but actually far more important, is the power to adjudicate over information. "Professional ecologists" will decide how much grazing is consistent with recovery of grasslands. Good—but which professional ecologists? Who decides between the conflicting opinions that invariably arise? Who decides who decides? Inevitably—but, I fear, invisibly to Deep Ecologists—not-doing becomes another kind of doing. In this particular case it might turn out to be good range management, but it would be neither simple nor passive.

The Far Green has many different subgroups, and some of these are far more willing to exercise power and even violence—as long as it is against people, and in defense of nature. Some advocate and practice acts of sabotage—such as "spiking" trees by driving huge nails into them to destroy the saws of lumberjacks. The late Edward Abbey's book *The Monkey Wrench Gang,* a story of ecosaboteurs, is a sort of Bible in those circles, and "monkey wrenchers" is a common term for those who go in for direct attacks against despoilers of nature. There is also a certain violence of thought and language, as in the article in the issue of *Earth First!* that cheered on AIDS as an antidote to human overpopulation.[12]

Although most mainstream environmentalists reject Earth First! and ecosabotage, there is no doubt that Far Green thinking has had a significant impact on the movement as a whole. It can be seen in the willingness of some major environmental organizations to follow Jeremy Rifkin in his crusade to drown biotechnology in a sea of sloppy thinking, in their hot campaign against the North American Free Trade Agreement and other such international (and obviously nonbioregional) commercial arrangements, and in the platforms of European Green parties that proclaim their allegiance to basic ideas of bioregionalism and Deep Ecology. The Far Green serves as an ideological albatross around the environmental movement's neck.

Its wider impact is less impressive, and also harder to gauge. In another proenvironmental critique of environmentalism, *Green Delusions,* geographer Martin Lewis concludes, "While radical views have come to dominate many environmental circles, their effect on the populace at large has been minimal."[13] This may be correct, but nevertheless I believe the Far Green needs to be taken seriously for several reasons—because it is a persistent influence in one of the most important political movements of our time, because it makes it easier for opponents of responsible environmental protection to dismiss all environmentalists as "ecofreaks," and because it fails utterly to engage in any realistic way the challenges the human species now confronts. By demonizing technology, it renders itself incapable of helping us to understand life in a high-technology, informatizing world; by enshrining homeostasis as a sacred value of ecology, it puts itself out of the running in relation to comprehending rapid change.

The Two Ecologies

There is a large—and growing—gap between the ecological ideas that inform radical environmentalism, and the ecological ideas of most professional ecologists. A new paradigm has emerged in ecological science: a shift away from the belief that the normal condition of nature is equilibrium. This new view—informed by chaos theory and bolstered by accumulating evidence on how organisms actually compete and coexist in ecological systems—is being advanced by scientists such as Daniel Botkin of the University of California at Santa Barbara. It is quite different from the more romantic sort of ecology, which still clings to the idea that ecosystems remain in a state of harmony and stability until they are disturbed by human action.

Botkin has pointed out that the latter kind of ecological theory has served as a subtext to environmental writing, all the way back to George Perkins Marsh. Marsh had written: "In countries untrodden by man, the proportions and relative position of land and water, the atmospheric precipitation and evaporation, the thermometric mean, and the distribution of vegetable and animal life, are subject to change only from geological influences so slow in their operation that the geographical conditions may be regarded as constant and immutable.[14]

This image of steady-state nature runs deep in much of environmental thinking: in the bioregionalist agenda of getting everybody to settle down and live peacefully within the timeless cycles of nature, in the Far Green's contempt for progress and change, in theories such as the "steady-state economics" of Herman Daly—the environmental movement's favorite economist—which are offered as a more "natural" way of thinking.[15] But, Botkin wrote, this very idea of nature is fundamentally at odds with more recent studies of how ecosystems actually work:

> Scientists know now that this view is wrong at local and regional levels—whether for the condor and the whooping crane, or for the farm and the forest woodland—that is, at the levels of populations and ecosystems. Change now appears to be intrinsic and natural at many scales of time and space in the biosphere. Nature changes over essentially all time scales, and in at least some cases these changes are necessary for the persistence of life, because life is adapted to them and depends on them.[16]

So we now have two ecologies: an ideological ecology that equates nature with stability, and a scientific ecology that equates it with change. The difference becomes a big difference as people become more engaged with proactive forms of environmentalism such as ecological restoration.

Restoration Rumbles

Not long ago hikers taking a Sierra Club field trip through the hills of Southern Illinois came across a crew of men busily burning and clear-cutting the forest. The hikers, all good conservationists, demanded to know what was going on. They found out that the tree-burners were also good conservationists, although of a different type. They were doing an ecological restoration—creating a living replica of the treeless, savannahlike areas called barrens that had once been common in that part of the country before settlers began putting out the naturally occurring fires and shooting the herds of buffalo, antelope, and elk whose feeding and trampling had helped prevent the region from becoming forest.

The hikers, without quite knowing it, had wandered into the bio-information society—caught a glimpse of ecological things to come.

Ecological restoration is a fast-growing and relatively new art and science. Restorationists build "natural" ecosystems—marshes, forests, deserts, and waterways more or less as they were at some period of time before human activities damaged them. These restored ecosystems, which can now be found in many parts of the world, are not precisely the same as those of decades or centuries ago; some species have become extinct and there are always a few stubborn exotics that can't be kept out. But they are infinitely closer to the pristine state than what you would get if you simply fenced off a territory and left it alone.

The oldest intentionally restored ecosystem is the Curtis Prairie in Wisconsin, where work began in 1934 under the supervision of ecologist Aldo Leopold. The ecologists set out to create a specimen of the great tall-grass prairies that had once stretched across millions of acres of the American heartland. Taking over an abandoned cornfield, they labored for decades to make it natural again. They hauled in tons of prairie soil, planted thousands of seeds and seedlings of native plants. And in time, under their careful cultivation, the abandoned cornfield became a stunning expanse of waving six-foot-high grasses and bright wildflowers.

Drawing on the lessons learned from that pioneer project, other restorationists have created more prairies. One of the most spectacular occupies 650 acres in the middle of the proton accelerator ring at the Fermi Laboratory in Illinois. That project established a couple of restoration firsts: the first use of heavy farm equipment—the same machines that other people are using to destroy native ecosystems—and the first attempt to restore wildlife species and habitat at the same time. Along with the tall grasses and flowers, populations of insects and ground squirrels, trumpeter swans and sandhill cranes were introduced into the new prairie.

Restoration projects can be found now in all kinds of ecosystems: forests and deserts, seashores and mountains. I have hiked into hills in Marin County, within commuting distance of San Francisco, where local volunteers are rebuilding eroded hillsides and reconstructing streams to bring back the salmon.

As ecological restoration grows, it accumulates its own body of skills and data—another kind of bio-information—and it also helps to raise fundamental yet disturbing questions about what, exactly, environmentalists ought to do (or not do) about the environment.

Some years ago I attended a conference entitled "Restoring the Earth"—the first major international gathering of professional restorationists. They came together, like professionals in any field, to get acquainted, network, and exchange information. There were panels and presentations on techniques and case histories, keynote speeches by world-class environmentalists such as David Brower and Stuart Udall.

On the whole it was a convivial and upbeat gathering, animated by a general feeling that the participants were all on the side of the angels. But here and there you heard a discouraging word, a note of concern that perhaps restoration was not really in the service of the environmental cause. Restoration is about very aggressively and deliberately *doing something* to ecosystems, and thus contrary to the "not do" ideology and its assumption that being an environmentalist means leaving things alone and/or getting other people to do the same. Some people grumbled that if this restorationist business went too far, it would just give people a license to destroy and pollute, and assume that the restorationists would come along later and fix it all up. These concerns are heard regularly now, a persistent troubled theme in the environmentalist world.

And the problem goes even deeper than the one that surfaced at that particular meeting, because the new ecology, and the bio-information that informs it, raises the question of what stage of an ecosystem's history should be restored. For most of us—and I fear that includes a great number of people now doing and advocating restoration—the answer is simple. You just, you know, restore nature. Make it like it *was*. But consider Botkin's reflections on the wilderness known as the Boundary Waters Canoe Area, along the Minnesota-Ontario border:

> Moose meander here . . . and people come every summer to canoe and hike and to discover wilderness. But what is the true character of this wilderness that they seek? The history of the vegetation holds some clues for us. That history has been reconstructed using all three kinds of evidence: written history; existing forests; and lake sediments. Pollen deposits from the Lake of the Clouds within the Boundary Waters Canoe Area indicate that the last glaciation was followed by a tundra period in which the ground was covered by low shrubs now characteristic of the far North, as well as reindeer moss and other lichens and lower plants.
>
> The tundra was replaced by a forest of spruce, species that are now found in the boreal forest of the North, where they dominate many areas of Alaska and Ontario. About 9,200 years ago the spruce forest was replaced by a forest of jack pine and red pine, trees characteristic of warmer and drier areas. Paper birch and alder immigrated into this forest about 8,300 years ago; white pine arrived about 7,000 years ago, and then there was a return to spruce, jack pine, and white pine, suggesting a cooling of the climate. Thus every thousand years a substantial change occurred in the vegetation of the forest, reflecting in part changes in the climate and in part the arrival of species that had been driven south during the ice age and were slowly returning.
>
> Which of these forests represented the natural state? If one's goal were to return the Boundary Waters Canoe Area to its natural condition, which of these forests would one choose? Each appears equally natural in the sense that each dominated the landscape for approximately 1,000 years, and each occupied the area at a time when the influence of human beings was nonexistent or slight.[17]

When you remind yourself that no restoration is ever perfect as a simulation of the ecosystem of any past period anyway, this rather basic piece of ecological wisdom does put the whole enterprise of ecological restoration in a different light. Perhaps we shouldn't even call it ecological restoration. Perhaps we should call it ecoconstruction or even ecoart. Perhaps we should call restored areas ecological theme parks. How about virtual nature?

The purpose of these comments is only to provoke some critical reflection, perhaps even a bit of genuinely deep ecology, about the meaning of restoration. I certainly do not want to disparage restoration itself, because it will be—by whatever name—an immensely important part of how we deal with the biosphere in the years ahead.

Thinking Big

In the late 1980s a San Francisco firm of landscape architects was awarded an interesting contract: landscaping Kuwait. It was the first time a team of professionals had ever been hired to remodel an entire country. The project was subsequently put on hold after the armies of Saddam Hussein invaded Kuwait and submitted it to their own rather more primitive notions of environmental transformation. But it is being revived, and people are beginning to propose others of similar scale. At the present time most restoration projects cover a few hundred acres at the most, but it will soon become possible to think about ecological restorations on the scale of, say, Cuba. Cuba is a country whose forests have been largely destroyed, and whose land has been historically dedicated to a single export crop—sugar—that is not likely to be able to compete with artificial sweeteners now being developed. The new techniques of agroforestry, which combine forest maintenance with various kinds of commercial productivity, will likely play a part in many such projects.

Such projects sound grandiose now—as daring and innovative as the idea of large-scale economic recovery seemed after World War II, until the Marshall Plan successfully breathed new life into the shattered economies of Western Europe. But they will become a familiar feature of the bio-information society. Restoration does not replace wilderness protection, any more than, in a postindustrial economy, information replaces industry. It just changes the picture, and expands the possibilities. And the restorationists are building up their own data banks and networks with information about tools and methods and the usefulness of different kinds of vegetation.

All kinds of reforestation and ecological first-aid activities are going on around the world now, and—although much of this is responsible work—not all of it quite meets the standards of dedicated restorationists. It sometimes leads to still more biological globe-trotting when, for example, fast-growing trees from South America are employed in reforestation projects thousands of miles away from their native bio-regions. To avoid criticism, some reclamation workers may use a sturdy exotic plant such as an African grass to stabilize soil at an early stage of a project and then replace it later with native vegetation. The Everglades project mentioned in the previous chapter is generally called a restoration, but clearly its object is not to transform central and south Florida back to some presettlement state of nature—hardly a likely prospect for a region that contains millions of people, plants and animals from all over, and the world's largest zoned farming area, the Everlades Agricultural Area. It's just a different approach to managing the region, an approach that seeks to correct some of the mistakes of past management. The planners call it "adaptive management," meaning that they acknowledge the need to learn by doing and the inevitability of errors along the way.[18] This is the philosophy that will guide the work now being undertaken; it is considerably different from the old Corps of Engineers mentality and also from that of environmental ideologues. It is a part of what environmentalism must become.

One More Shade of Green: Proactive Environmentalism

We are living in a time of rampant mobility, huge cities, exponentially growing human populations and burgeoning, converging technologies—and such a time desperately needs a vision that is able and willing to engage its realities.

The moment has come for a new environmentalism. Those who embrace it must be no less concerned about the inherent value of other living things, the dignity of indigenous peoples, or the restorative powers of the great outdoors than other environmentalists have been. They must be equally willing to do battle against the crass antienvironmentalism that would roll back all protections of clean air and water. But they must be much more willing to face the rapidly changing conditions of our time and to take an active part in the creation of a global bio-information society.

The folly of the antienvironmentalist right lies in its ideologically driven resistance to all evidence that the human species and its civilization have impacts upon the biosphere. The folly of the bio-regionalist left lies in its eagerness to demonize human civilization and to believe that we can or should live in the world without affecting it. The job for the rest of us is to keep learning about those impacts and finding patterns of management and governance—ways to live in the world with a full willingness to accept the extent of human power to change it—that are sensible and humane.

A proactive environmentalism must be globalist, activist, future oriented rather than past oriented, supportive of economic development, more interested in learning from past mistakes than in harping on the wrongheadedness of those who made them, and capable of understanding and using new technologies. There are already signs of this: many people are already making considerable progress toward a "green biotech" for developing better methods of agriculture, recycling, and pollution control. The Ecological Society of America has gone on record in favor of continuing research and development of biotechnology, and several of its leaders have expressed enthusiastic support for "the potential of biotechnology to provide ecologically sound alternatives to some current practices in managed ecosystems."[19]

Although this kind of environmentalism has not yet captured the popular imagination, it is already here in practice. Many working ecologists and environmental activists are dealing with real problems in a much more pragmatic, proactive, and frankly creative fashion. They are doing so because it turns out that no matter how much we liked *Pocahontas,* we really don't know how to leave nature alone.

We need to face a central truth about the bio-information society that is growing up all around us: the world is becoming more (not less) anthropocentric. Human powers and human responsibilities increase with every research project, with every deposit added to the gene banks and the data bases. We are moving speedily toward a managed biosphere. It may well be mismanaged, it will probably be managed in different and even conflicting ways by different people in different places—but it will not be left alone. As human populations continue to grow, urbanize, and become more technologically proficient, stresses on many of the world's ecosystems will increase. So will human ability to identify the problems and invent responses. There will be many kinds of responses, many management alternatives to consider. Wilderness protection is one alternative, and often a very good one. But it is still management.

I have gone hiking many times in the High Sierra, usually in the John Muir Wilderness area. It is a wonderland of granite cliffs and clear lakes and fields of wildflowers, a peaceful place where people come to enjoy the rugged environment, watch the soft alpenglow at dusk, lie in their sleeping bags at night contemplating the stars and satellites. It is wilderness, and it is managed. The forest rangers maintain the trails and ask you not to cut corners. They politely write you a stiff ticket if you camp too close to a lake. Sometimes bears that get out of hand are sedated and hauled away by helicopter to more remote regions. In some wilderness areas (such as those national parks in Costa Rica) management involves periodically throwing out hungry squatters who surreptitiously invade to carve out small plots of farmland. Management always involves tradeoffs and differences of opinion, always comes with a price tag. Ecology is politics.

One reason we can't leave nature alone—one of many—is that everybody keeps *studying* it. Wherever information goes, management follows. People test the water and the soils and the air, count the birds and beasts and flowers—and the information leads to actions. We have become accustomed to water-quality and air-quality regulation and wildlife protection on a local and regional scale, and the accumulation of new bio-information is leading quickly toward management on a global scale.

Science, Ecology, Politics, and Art

Disputes about ecological management get messy, and the disputants have a hard time figuring out how to settle arguments. Some invoke science, but the "facts" are never complete and always questionable. Others invoke nature, an equally shaky tactic. Everybody searches madly for some nonhuman principle to use as a final arbiter and authority.

On all sides, people resist acknowledging something that should be obvious—which is that in all environmental disputes there is a heavy element of personal taste. Subjective—and *human*—perceptions, emotions, and values are always present. Ecology is art as well as science, and ecological disputes often boil down to competing ideas of beauty. Most of us say we love the experience of nature, but some people would love nature in an Illinois barren *with* trees and some people would love it more *without*. Different groups of human beings make up arguments (all commendably nonanthropocentric) for the management alternatives they prefer.

As we move further into a bio-information society, people will come to understand the inevitability of management and the inseparability of human aesthetics from ecological politics. As they do, they will have some very interesting new arguments—and they will create not only new environments, but also a new environmentalism, a way of thinking about such matters that will be quite different from the sensibility of our own transitional and rather confused time.

CHAPTER TWELVE

The Infinite Schoolhouse

No matter how far we go into the future, there will always be new things happening, new information coming in, new worlds to explore, a constantly expanding domain of life, consciousness, and memory.

—Freeman Dyson[1]

What is evolution, if you work your way down to the essential heart and soul of it? What is it *about*? Certainly it is about those old Darwinian standards, adaptation and survival. It is also about change—and definitely not just the gradual change that people sometimes have in mind when they distinguish between evolution and revolution. But most of all, it seems to be about learning—acquiring, using, and communicating information.

Everything that lives is engaged in learning, and as the strange new science of artificial life unfolds it begins to look like the little electronic entities inside the computers are learning too. Species learn the hard way—through natural selection. Individual animals learn from their genes and also learn from experience during their lifetimes. Roughly speaking, the higher the animal on the evolutionary scale, the more it relies on the information it acquires through experiential learning and the less on the information it was born with. Learning is universal, present in all action and all life. Human beings learn how to control disease through antibiotics, bacteria learn how to resist antibiotics, human beings learn they have to learn new ways to fight disease. A computer, with the right software, learns how to play a better game of chess. So it goes around the world and, we have every reason to suspect, around the cosmos. Even God is learning, according to the process theologians. Alfred North Whitehead's followers describe God not as the famous unmoved prime mover, but rather as a living, changing force that participates in evolution as it unfolds.[2] The opposing, nontheistic version of cosmic learning is summed up nicely in Richard Dawkins's famous metaphor of the blind watchmaker: learning just happens. "Natural selection is the blind watchmaker," Dawkins claims, "blind because it does not see ahead, does not plan consequences, has no purpose in view."[3]

The Dance of the Three Lineages

When I referred to learning by bacteria and humans and machines I was mixing together three different kinds of learning. These appear at different stages of evolutionary development and use different information systems. The first, the biological learning system—which appeared at a very primitive stage of life on Earth, yet a very advanced stage of the evolution of the cosmos—is driven by the rules of natural selection, and uses the genes to store and communicate information. The second is culture, the "cultural DNA" of signs and language, by which we become theoretically capable of learning from anybody and communicating information to anybody. And the third is the "exosomatic" (that is, outside the body) system of devices. This includes all the artifacts—from cave drawings to books to telephones to the huge system of satellites, cables, computers, and data banks now blossoming around the Earth—that are not parts of our biological bodies but yet become integral parts of our thinking, learning, and communicating. Susantha Goonatilake of Sri Lanka and the United States, one of the most interesting of the current generation of evolutionary theorists, calls these three systems "lineages"—a nice word that suggests a heritage, something continuing yet changing over time like a royal family.[4]

Three different information systems, then: genetic, cultural, exosomatic. Each grows out of the previous evolutionary stage and each, in its own way, advances the general progress of the species. I should make it clear that when I talk about information here I am using the word as classical evolutionary biologists have used it—to mean instructions on how to deal with the environment. And when evolutionary biologists talk about progress, they mean not just progress in adaptation, but progress in adaptability.[5] Adaptability has to do with responding to change, which is something different species do in different ways with different degrees of success. *Homo sapiens* is famously adaptable, and human adaptability is the product of cultural evolution. Our bodies are not all that adaptable by themselves, and your instincts alone would not get you very far if you were suddenly stranded in a hostile climate.

The evolutionary transitions that we have discussed in previous chapters involve not only learning, but learning new ways to learn.

Periodically—as in the evolution of *Homo sapiens*—the process takes a leap to another stage, begins a new lineage, and then moves along new paths of exploration, growth, complexity, and adaptability. Nontheistic evolutionists such as Dawkins' colleague Daniel Dennett hasten to remind us that such evolutions of evolution are engineered by "cranes" and not by "skyhooks"—meaning they are inventions that evolution constructs and then uses to lift itself to a new level, not helping hands extended down from above.[6] But whether you go in for cranes or skyhooks, you are confronted with the rather obvious yet mind-boggling, inspiring, and occasionally frightening fact that the evolutionary process occasionally learns a new way to learn—and, when it does, changes its rate of change.

Goonatilake points out something most strange and fascinating about the three lineages of information, which is that each has its own evolutionary speed; each is more adaptable than the one that came before it. Culture changes more quickly than our genes, and our computers are now changing more quickly still.[7] And something else happens: each new lineage transforms the one that preceded it and created it. When culture appeared, human biological evolution changed—most notably by favoring individuals with larger brains, greater ability to process the symbols that are the stuff of culture. The appearance of each new information system—language, writing, movable type—produced a new culture, and new kinds of cultural change. The global information-communications networks are now creating a global popular culture, enabling people to process symbols in new ways, and affecting all local cultures.

More yet: we have been talking about the transformative leaps in *human* evolution that are brought about by the emergence of new information systems. But obviously it was not only the human species whose evolutionary fortunes were affected when speech and writing made their appearances. Culture in human beings came to include knowledge of how to breed plants and animals, and how to modify the Earth through farming and mining and canal building. And the present acceleration of information-systems growth is bringing with it a host of new rule changes in the biological evolution of species and the ecological workings of the planet itself.

One final point is that these information lineages are not really separate. They appear at different evolutionary stages, they have different characteristics that we can analyze and measure and describe—but in practice they get most intricately and creatively mixed together. They converge and interpenetrate in a multitude of ways. We opened this book with an examination of how one kind of human cultural learning—

the one we now call biology—evolved along with information and communications technology. New tools and technologies transform culture and culture transforms genetics—and it also works the other way around: genetic influences—as the sociobiologists tell us—help shape culture, and it was only through culture that people were able to make the new tools and technologies. A rich and complex evolutionary dance, and it is an utter waste of time to try to reduce it to some stale rubric such as "technological determinism."

Evolution now involves all three of these learning systems, and we can make no sense of it at all—which means that we cannot understand what is happening to us, and to our world—without taking these into account. And that means, especially, recognizing that an explosion of learning in the third lineage, the exosomatic system, is changing everything.

Paradoxical as it may seem now, on the threshold of a century that will be dominated by advances in genetic science, evolution in the human species is not, at the present time, really a matter of genetic change. We are still born with essentially the same brains and bodies as our ancestors of thousands of years past. We then proceed to become vastly different from them as we take in the cultural information of our time, as our brains and bodies are augmented by inventions they could not have imagined, as we plug ourselves into new connections with other people and with the biosphere.

At the moment, the most spectacular genetic change taking place in the human species is probably the one I mentioned in Chapter 7, the sharp reduction of inbreeding that results from increasing human mobility. The vast human migrations of our time may present many cultural and political problems—and such mobility certainly doesn't please the bioregionalists—but from a genetic point of view it is a healthy development: something like a large-scale equivalent of importing Texas cougars to beef up a diminished Florida panthers gene pool. There is also strong evidence to support fears that some deterioration in the gene pool results from the cumulative effect of all the things done to cure diseases and save lives. The human genome is probably fluctuating in many ways, but no fundamental *genetic* change in the human condition appears to be in the offing. As far as the DNA knows, *Homo sapiens* is still *Homo sapiens.*

Perhaps further changes in the human gene pool *will* result if germline therapy becomes a practical (and socially accepted) reality. Conceivably some genetic defects will greatly diminish or even disappear. But this is not the direction in which I would look for clues about the near-term future of human evolution. I do not anticipate the appearance of smarter brains and stronger bodies through genetic wizardry; maybe someday, but not soon. What I do anticipate are smarter computers, more fully integrated into our lives in countless ways, and further augmentations of the human body. In fact I suspect that computers will change so quickly that they will more or less disappear—cease to hold the conspicuous position they currently occupy in our desks, laps, and conversations and be all around us in the form of smart tools, smart houses, possibly even smart clothes. The people at the MIT Media Lab have been promoting the concept of the "Bodynet"—the human body in a sense wearing its computer and connected through it to other people or information systems. In these and other ways, the human species will evolve through the lineages of culture and technological change.

There has never been an explosion of advances in information systems comparable to the one that is happening now, and it would take a very foolish futurist to pretend to know exactly what forms it may take. We cannot predict that—but we *can* safely predict that it will continue for some time to come. It is not even particularly risky to hazard a guess that its rate of change will continue to increase. It will produce not only augmentations of the human mind and body, but also more means of managing other species and ecosystems, a faster and much more elaborate global communications web, and increasing ability to study the planet as a whole and evaluate feedback from its various systems.

So the framework for considering evolution now is: minimal change in the genetic nature of the human species as a whole, but exponential change in the exosomatic information systems—and, inevitably, very rapid changes in culture. Technological change will force cultural change to take place, but will not determine its outcomes. What that means is that people will have to make more choices.

Culture is the part of evolution that is the closest to our consciousness—because our consciousness *is* culture—and it is also in some ways the most mysterious. We have seen an incredible amount of cultural change in the past few decades, and will see even more in the decades to come. In fact, I expect the period just ahead to bring the most rapid cultural changes the human species has ever experienced, and to test our capacity for such change to the limit. People frequently make pronouncements about how much cultural change is possible, but there are no rock-hard rules for that—or, if there are, we don't know them. This period will bring a continuing globalization of culture, with increasing communications and mobility, and it will bring countermovements—fundamentalism and localism and conservatism and backlashes of many kinds that resist the globalizing trend.

In these closing pages I want to consider three sets of cultural issues that are of great importance as we move into this global bio-information society: ethics, equity, and governance. And finally, I will touch on what I think is the most important cultural question of all—what story we tell, how we talk about what we are talking about.

Ethics in Evolution

One of the commonplace observations about the proliferating bio-information technologies is that they present people with new ethical choices. That's quite true, but there is more to it than merely making ethical choices. We also have to make choices about whose ethics we are going to use when we make choices.

We are now taking up residence in a postmodern global civilization wherein the ironclad values and beliefs of traditional religions, social orders, and ideologies are being revised on all sides. Reality, I have argued, isn't what it used to be.[8] You may decide to make your own life more manageable, and reduce the range of choice to less formidable dimensions, by declaring allegiance to an organized belief system such as the Roman Catholic Church—but that is yet another choice. If you make that choice, you have to keep making further choices about whether to stray from the flock or keep the faith, about whether to take the whole package or select the values and beliefs that work best for you. Many people, to the distress of the Church fathers, take the latter path: they become what are known as "cafeteria Catholics," picking and choosing among the doctrines. We also have cafeteria Protestants, cafeteria Muslims, cafeteria Marxists, cafeteria old-fashioned Americans.

The job of contemporary ethicists, then, is much different from that of traditional authorities who could simply lay down the law. Ethicists now are charged with reinterpreting and reinventing old systems of value and belief, helping people find their way into a new world of multiple and ever-changing realities. For most of us, ethical principles relevant to the present age are not reliably located in the sacred works that were written down several centuries—if not several millennia—in the past. We may get some guidance from those ancient sources, but the fact is that ethics are being created, constructed and chosen—and negotiated—in our time. They have to be negotiated, because there are so many different possible positions—indeed, fundamentally different ethical approaches—to any issue.

Understandably, people often try to take a shortcut through this confusion by finding some simple ethical key that seems to be clear and acceptable to everybody. Nature is one of the favorites—the most attractive guideline to grasp, and probably the least useful. The trouble with that particular basic principle is that we believe it has a solid, universal, cross-cultural meaning, and thus relieves us of some of the burden of making choices. But it doesn't: ideas about what is nature, or what is natural behavior, are enormously divergent and quite culture-specific. Anthropologists have shown that what one people regards as healthy, natural behavior—an occasional spot of head-hunting, for example, among the Ilongot tribesmen of the northern Philippines, or oral sex between male Sambian children and adolescents in New Guinea—is shockingly depraved to another.[9] It is not at all clear that all cultures even have a concept of nature comparable to our own.

If what you mean by nature is some patch of the Earth's surface totally unaffected by human agency, forget it. There is no such ecosystem, no such animal, no such place—and hasn't been for quite a bit longer than Bill McKibben, the author of the recent book *The End of Nature,* suspects.[10] If what you mean by nature is the cosmos, fine; but that of course includes us, our machines, our sewers, and slag heaps. Nature is a perfectly good and handy word, no more ambiguous than many others in our ambiguous vocabulary, for talking about wilderness areas, or the love of open spaces and living creatures, or things we do without giving them much thought—but it has no precision, and no value whatever as a guideline to right and wrong actions. It is somewhat comparable to our common use of the terms "black" and "white" to distinguish between races—that is, generally unavoidable and fundamentally meaningless.

It's quite understandable that some people grasp at nature—the way others grasp at scientific objectivity, and still others grasp at religious doctrine—in hopes that it will offer some extrahuman, eternal guidelines through the complexities of our times, serve to shut up wrong thinkers and settle arguments once and for all. But it doesn't. And neither does science, and neither does religion. Each is useful, in limited ways, but none is sufficient. None rescues us from the blessing and scourge of our time, which is a multitude of options.

If you are looking for an insight into the ethical situation of our times, go to any medium-sized public library, and you will find it easily. Locate the section on ethics, and you will see books of right-wing ethics and left-wing ethics, religious ethics and secular ethics, feminist ethics, gay ethics, New Age ethics, back-to-nature ethics and three-cheers-for-science utilitarian ethics. The authors of some of these books seem to be willing to admit frankly that they are aiming at a limited audience—a certain subculture—but a lot of them seem to think they can speak to and for everybody. What you need to do is step back a pace or two, regard those shelves as a whole, and you will see the truth about the world we live in: a world of many ethics.

This is not by any means a hopeless situation, although some find it so. It is, rather, a time of great vitality. Ethical answers do not get handed down to us by Gaia or by God and they do not get printed out of our computers. We find them in dialogue—perhaps multilogue is a better word—with other people, in dialogue with the traditions of the past, and with ever-changing information. In a sense we create them, but we do not create them capriciously, out of nothing. We create them by learning, and also by unlearning—periodically revising our information and assumptions. And when we make ethical choices—which are frequently life and death choices—we must make them with a certain humility, knowing that our information is incomplete.

The best example of the kind of ethical process we need more of in the bio-information society is provided by genetic counselors, who work with people as they make decisions about such matters as genetic screening. The counselor provides the client—often a couple of prospective parents—with scientific information, explaining the findings and the available alternatives. He or she also encourages them to explore and consider the teachings of whatever cultural or religious tradition they may belong to, in order to see what guidance it offers them. And the counselor also encourages them to talk with one another, with family, with other people who have faced similar problems. It is a human process and embodies a certain humane wisdom, and is far superior to hitting people over the head with the Bible or the latest work on Deep Ecology. Ethical discussion in our time is either dialogue or it is nothing.

Equity in Evolution

The great American political scientist Harold Lasswell defined politics as a matter of who gets what, when, how.[11] It's a rather unromantic (and of course anthropocentric) definition of the subject matter of Plato and Montesquieu and Marx, the occupation of Caesar and Lincoln and Churchill. But it has always made sense, and I am beginning to think that now—even now, when political actions can have evolutionary implications, affecting not only all people but all life on Earth—it makes more sense than ever.

The most pressing, explosive, unavoidable issues that societies face today are questions of equity. Matters of who gets what, when, how. The enormous gap of wealth and opportunity that has opened up in the world may not be a single "global problem" comparable to the threat of global climate change or species extinction—but it is the problem that can block solution to all the others because it renders people who need firewood unreceptive to warnings about CO_2 in the atmosphere, drives people who are hungry or poor to hunt down endangered animals.

And the great unanswered question of our time is whether the information revolution and the biological revolution will serve to close that gap, or to widen it. I have participated in various conferences and conversations on this issue, and they generally are divided into two groups. Some people, the cyber-optimists, are convinced that information and biotechnology do not operate by the old rules of resource scarcity, and will diffuse quickly around the world—bringing new solutions to problems of poverty and disease and environmental destruction. The cyber-pessimists are sure that we will get only a widening of the gap and perhaps even—as the wealthy tekkies skip merrily onward into the bio-information society while the unenlightened masses remain mired in polluted and overcrowded misery—a split more deep and ugly than the world has ever known.

I am quite sure that the pessimists are right on one point: which is that the information revolution has the ability to make the gap more visible, painful, and dangerous. We in the developed world may not do so well at providing food, housing, clothing, and jobs to people in the poorer regions of the world, or enabling them to provide for themselves—but we seem to be brilliantly adept at exporting images, sometimes vividly framed in television and filmed dramas, of how wealthy and well-fed people live.

Other, more complex kinds of information diffusion do seem to be taking place. People everywhere are now beginning to understand that there are different kinds of resources in the world—including resources such as genes and biological data that for all practical purposes didn't exist until fairly recently—and that these can have a very real life-and-death value. Bio-information issues are complex, but some basics are easy enough to grasp. Informed people live longer and healthier lives. Informed farmers grow more and better food. Informed communities do a better job of managing their environments. Informed companies make more money. A little harder to grasp, but still being widely discovered now, is that to be "informed" means not only to have information, but to know how to use it and—most important of all—to know how to keep learning.

The global information inequity is not exactly a North–South gap, because what some people call "critical transnational communities" of advanced research and development are already blossoming in the most unlikely places—or, to be more precise, blossoming without reference to place at all but instead forming as geography-free networks. Yet, even though the geographic picture is becoming increasingly murky, it is still clear that at the present time the benefits of the information age flow more copiously into the lives of people in the developed nations.

Can this be remedied by massive transfusions of the new resources into the economies of underdeveloped nations? Jean-Jacques Servan-Schreiber (international development activist and former director of the World Center for Informatics and Human Resources) thinks so; he has called for a "global transfer of information" to help people in the Third World improve their farming and health care, develop sustainable industries, and progress toward membership in a global information/service economy.[12]

Others are, for various reasons, highly unenthusiastic about this idea. People on the political right tend to worry about how such transfers might harm businesses and industries whose stocks in trade are high technology and information, who have invested heavily in research and development and are entitled to a return on the investment. People on the left worry about how they might disrupt the cultures and social structures of those who are supposed to be the beneficiaries. Some think such transfers are impossible because the educational and scientific institutions in the Third World aren't adequate. Some think the momentum of the information and biological revolutions is so great that the transfers will happen whether anybody makes them happen or not.

The scientific discussions about the future of the potential superdrug interleukin-12 offer one good example of who actually gets what, when, and how in an industry driven by the demands of the marketplace. IL-12 appears to have applications against an astonishingly wide range of diseases, from cancer and AIDS to Third World scourges such as malaria and leishmaniasis. Tests against many diseases are under way, with promising results at the early stages, but representatives of the companies that hold the patents on IL-12 recoiled in horror from the possibility that they might have to develop a product for the Third World countries first. A *Science* writer, covering a National Institutes of Health symposium on IL-12, reported that a representative of one of the involved companies, Genetics Institute, Inc., had warned that "the success of IL-12 against, say, leishmaniasis at this point would be a 'disaster' for the company; G.I. would be committed to scale up manufacturing costing millions of dollars and would probably wind up distributing the drug through the World Health Organization, which would in essence give it away, leaving the company with huge costs and little or no revenues."[13]

You can hardly expect companies dependent on large infusions of capital—companies that often go for years, spending millions of dollars on research, before they produce a saleable product—to be uninterested in returning something to their investors. But the profit motive inevitably leads to much more scientific research being aimed at developing products for people with money. Sometimes there are trickle-down effects—eventually products become more widely available, prices drop—but in the interim many people die, and many people suffer. Attempts to fill the gap include research supported by international organizations such as UNESCO or by development-aid agencies of various countries, and private funding organizations such as the Rockefeller Foundation. All of these deserve much more attention and public support than they get.

Sustainable development, the most all-around popular slogan of our time, is one way of talking about the equity issue. And criticisms of it basically have to do with what kind of equity we need to be concerned about. The sustainable development people are mostly concerned about intergenerational equity—being careful not to plunder the resources that may be needed by people in the future. But the British economist Wilfred Beckerman, in another recent antienvironmentalist broadside—this one entitled *Small Is Stupid*—argues that the whole sustainability movement is only putting the needs of future generations (about which we know nothing) above the wealth and welfare of those who are alive now.[14] What we should be doing, he argues, is searching for ways to improve the distribution of resources among the current world population.

Both ways of thinking about equity—worries about future generations and worries about present gaps between haves and have-nots—flourish in the information society, and follow inevitably from the globalization of media and politics. The information age brings a sudden, rude, and shocking expansion of horizons. The wealthy and fortunate turn on their TV sets and see famines in Somalia. Meanwhile, starving people in Somalia get the message, often rather vividly, that a lot of people in the rest of the world are sitting around in restaurants worrying about which wine to order. Animal rights activists do their best to make us *feel* the suffering of other creatures. Scenarios of global disaster rub our noses in the future, fill us with guilty concerns about what we are doing to our descendants. The political landscape in which we live stretches in all directions, expands in space and even time.

Toward an Ecology of Governance

Sooner or later, most of the practical evolutionary issues of our time —all the questions about such things as genetic therapy, organ transplants, transgenic animals, biotechnology, gene banks, ecosystem management, sustainable development—become questions of governance. This is a difficult matter to comprehend, because it is a subject that many of the people thinking about the information revolution don't take very seriously. They seem to regard the usual institutions and issues of politics as kind of quaint. Kevin Kelley, for example, entitled his interesting book about the information revolution *Out of Control,* and concludes it with a statement that governance will be only in the form of control from the bottom up:

> When everything is connected to everything in a distributed network, everything happens at once. When everything happens at once, wide and fast moving problems simply route around any central authority. Therefore overall governance must arise from the most humble interdependent acts done locally in parallel, and not from a central command. A mob *can* steer itself, and in the territory of rapid, massive and heterogeneous change, only a mob can steer. To get something from nothing, control must rest at the bottom within simplicity.[15]

Here we have the concept of self-organization again and, as I have indicated at several places, I think it needs to be taken seriously in counterbalance to more traditional ideas of hierarchy and top-down control. Anything as complex as today's world is not going to be governed out of a central command-post, like the one I saw in Biosphere II. (As it turned out, that didn't even work too well for Biosphere II.) But the present global situation is not simply one of spontaneous bottom-up decision making. As the world organizes itself into a global civilization with growing governance responsibilities over the biosphere itself—which it is now in the process of doing—it both centralizes and decentralizes. It comes together and spins apart at the same time.

For all the talk of downsizing government and self-governing systems, there is probably more formal governance and politics in the world now than there has ever been. There is centralized power, there is authority, there are hierarchies. There are, to begin with, more nation-states. These aren't the sovereign fortresses they once were (or once believed themselves to be), but they still do a lot of governing: they pass and enforce laws, collect taxes, raise armies, regulate businesses. They create and use international organizations such as the United Nations and the World Trade Organization. They also create all kinds of international "regimes" through such mechanisms as the Law of the Seas treaty. The nation-states are still alive and well, in no danger of wasting away as some of our more adventurous futurists have thought they might.

But they do not have a monopoly on governance. Deliberation, decision-making, and even policymaking go on at many levels, in many places, involving many kinds of players. It has been obvious for some decades now that transnational corporations rival the nation-states as governing institutions. More recently, observers of global politics have been struck by the astonishing growth of nongovernmental organizations (NGOs) concerned with such matters as population, the environment, human rights, women's issues—growth in numbers, growth in size, growth in influence. There was a time when international conferences were gatherings of politicians and diplomats; now they tend to be global town meetings—gatherings of public-interest activists and other kinds of nongovernmental players. A UN official told me that over 1,200 NGOs were accredited to the Cairo conference on population. That's the number of *organizations,* most of whom brought several people. NGOs are deeply involved in policymaking, at all levels of governance—and often they have not only influence, but power. In some developing countries, local NGOs are recognized as more important than the local governments. This last may be bottom-up governance, but in practice it is often made possible by top-down financing from foundations, church organizations, businesses—and governments. At the Copenhagen "social summit" in 1995, Vice President Al Gore announced that the Clinton Administration would begin to channel nearly half its foreign aid through private organizations—i.e., NGOs—rather than through governments.

Then there are other institutions that can really only be described as networks—that don't have capitols, flags, constitutions, or even leaders in the traditional sense, but are nevertheless parts of the global governance system. The best example is the international currency market, which every day pushes about a trillion dollars' worth of money around the world. This huge global electronic system was created not to affect policy, only to buy and sell—but it has become a *de facto* agency helping to determine the value of currencies. The fact is that governments do not have complete control over their currencies, and neither does any formal international organization or alliance of states. As Walter Wriston put it in his book *The Twilight of Sovereignty*, all governments are now on the "information standard," just as most were once on the gold standard:

> Markets are voting machines; they function by taking referenda. In the new world money market, for example, currency values are now decided by a constant referendum of thousands of currency traders in hundreds of trading rooms around the globe, all connected to each other by a vast electronic network giving each trader instant access to information about any factor that might affect values. This constant referendum makes it much harder for central banks and governments to manipulate currency values.[16]

Information systems of all kinds are now parts of the global governance system. The global news media do far more than just report from the sidelines. They can bring distant events into our living rooms, create villains and heroes, influence which issues we become concerned about and which ones we neglect. One study of the growing role of "media events" itemizes three kinds of events—contests, conquests, and coronations—that are "scripted" for public consumption: many of these are now global events. By *contests* the authors of the study mean events such as the Olympics. The category of *conquests* includes heroic events such as the first moon walk and real battles such as the Iraqi War. *Coronations* are big-time ceremonies such as the wedding of England's Prince Charles, which was, at the time, the most-watched event in history.[17] Other kinds of events—natural disasters, terrorist acts, lurid crimes, political scandals—explode out of their local contexts and become, for a time, everybody's business. These events are, in a sense, creators of global culture. They also form the theater of global politics. The career prospects are not brilliant these days for a political leader who does not have a flair for acting on this stage—or, better yet, doing a bit of directing and scriptwriting.

The parts of the global information system that political scientists generally pay little attention to are the ones that I believe we will come to recognize as the most important of all—the growing and increasingly interconnected bio-information networks of remote sensors, researchers, data banks, gene banks, monitors of the Earth's oceans and air and land spaces and animal populations. These will do more to set the agenda for the twenty-first century than many of the more traditional movers and shakers of world politics—because they are truly learning systems, and will keep forcing us into greater awareness that human governance is inseparable from the life of the biosphere.

The phrase "new world order" has ceased to have much meaning—particularly in view of the failure of international institutions to establish the post–Cold War era of world peace that the term implied—but a new international system of governance is being created. This is happening not in one single act comparable to the founding of the United Nations a half-century ago but rather—the self-organization vision seems more appropriate here—in a multitude of separate yet interconnected acts of creation. The system includes the nation-states, the formal international organizations and regimes, the multinational businesses, nongovernmental organizations of all kinds, networks such as the international currency market, the global news and entertainment media, and the great world-encircling noosphere of bio-information networks.

Over the past few years the U.S.–Canadian Meridian Institute and other associated organizations have sponsored several conferences on the subject of "global governance," bringing together people of various disciplines and persuasions in an attempt to make sense of this. Our thinking gradually coalesced around the proposition that there are three different and competing visions of what is happening and what needs to happen. The first is the *state-centered* vision of a world governed—much as it has been in past centuries—by sovereign nation-states acting in the name of national interest, ruling through laws and treaties and power politics. Former secretary of state Henry Kissinger's book *Diplomacy* is an eloquent expression of this worldview.[18] The second is the *world-centered* vision of creating a new super-sovereign government, either by expanding the UN or—as some world federalists would prefer—writing a global constitution and founding a new order from scratch. The third is the *multicentric* vision of many organizations and kinds of organizations, overlapping and interpenetrating, with decisions made in many places. Different participants in these conferences had different names for such a system. Political scientist James Rosenau called it "polyarchy."[19] Diplomat-futurist Harlan Cleveland called it a "nobody in general charge" world.[20] Anthropologist Mary Catherine Bateson called it "an ambiguous world order."[21] I preferred the term "ecology of governance," which describes a dynamic, ecosystemlike mélange of organizations and individuals—and definitely not a stable one, because all the parts keep changing and so does the system as a whole. It has some of the qualities of self-organizing systems, but it is not all bottom-up: among its parts are many players—such as the Roman Catholic Church, autocratic nation-states, committees in the U.S. Congress, many corporations and private organizations, and armies everywhere—that are governed by rather more old-fashioned ideas of authority.

In Search of a Story

If I were to summarize the various evolutionary processes we have discussed, pull them together into a single statement about what is happening to the human species, I would say that we are seeing a shift of the boundary between the given and the made.

Throughout history people have known a dividing-line between the given world and the made world: between the areas of life that they understood to be simply handed to them—by nature, by tradition, by God, by force of circumstance—and those that they could to some extent shape according to their own needs and desires. The great majority of people have lived in societies almost entirely dominated by the given, and accepted both the comfort and the confinement that come with such an existence. But here and there—in Europe during the Renaissance, for example—the boundary moved, and people were gripped by an exhilarating sense of possibilities. This comes through clearly in the work of Renaissance writers such as Cellini, Machiavelli, and Pico della Mirandola. They felt life was theirs to seize upon and *make*, and out of that feeling poured great creativity in art, philosophy, governance.

Today we have possibilities that put the Renaissance to shame.

We have seen in recent history—and are seeing now—a sudden, dramatic shift in that ancient boundary, an enormous expansion of the life space in which we can (and often must) choose, create, construct. This shift is propelled along by information and technological change, but the increasingly intimate connection between our technologies and our organic lives does not diminish human freedom or power. The effect is precisely the opposite, and it transforms both personal and political existence. Birth control technologies empower us to make choices about who is to begin life, while "right to die" options empower us to make choices about when and how we will end it. People create and manage ecosystems, and the human species is rapidly becoming responsible for the management of the entire biosphere. People even make conscious choices about what they believe to be true and how much they believe it, what values they hold and how strongly they hold them, what rituals they practice, which social groups they belong to and are influenced by. Some of these changes have been working through the world for centuries, but nothing in history compares to the increase that has taken place in our time in human ability to shape the physical and cultural environment.

This is of course most apparent in the advanced industrialized societies. Like everything else in the world, the new freedoms and powers (and the special worries that come with them) are inequitably distributed. But those inequities should not lure us into the belief that the boundary shift is occurring only in the places where technology and pluralism are most advanced. If you believe that, you are likely to misunderstand what is happening to remote and primitive peoples as they struggle to manage their environments and re-create their traditional cultures—to make their worlds, in short. It is happening to all people.

Whether we like it or not, we are all worldmakers. And many of us *don't* like it. We look about at the hideous and harmful environmental damages that have been caused by exercises of human power over nature, at the painful stresses that come with new freedom to shape personal and social life, and we wonder if there might not be some way to put evolution into reverse gear.

Reverse-gear movements are an important part of contemporary politics. Every developed country has its forces of reaction—usually appearing in two quite different shapes and taking up arms against different evolutionary lineages. On the right are the reactionaries against cultural change, who long for a fantasized golden age of tribal or national purity. On the left are the reactionaries against technological change, who long for a fantasized golden age of environmental stability. But even as people desperately play the psychological and political games that psychoanalyst Erich Fromm aptly called escapes from freedom, we never really escape our peculiar fate.

With increasing bio-information, people are being forced to recognize that they can and do alter the living physical environments that surround them. They are also—because they live in mobile and pluralistic societies, amid the clash of cultures and countercultures—becoming more aware of their ability to participate in constructing the symbolic environment. They see that the invisible but all-important webs of value and belief and meaning that give shape to their lives, and the formal institutions of governance that structure their societies, are also the products of human imagination—made by people, and subject to being remade by other people.

Only now, as the century draws to a close, are we able to put the pieces together and—for the first time in history—catch a breathtaking glimpse of the transition that has been forming and gathering speed throughout the modern age. It is a transition into a world in which people become responsible for the physical and symbolic environments in which they live: in which neither ecology nor culture—neither nature nor nurture—simply happen to them. This transition is perhaps the most difficult ever undertaken—and hardly made any easier by our dawning awareness that we *are* undertaking it.

Completing it will be the dominant activity of the 1990s and the twenty-first century.

For many people the work will be done mostly at the local level. It will involve the making and remaking of small communities and regional ecosystems, perhaps nothing more than maintaining a home, a family, or a life style. But nobody will be permitted the luxury of separation from larger systems, particularly global ones. When the late biologist Rene Dubos coined the phrase "act locally, think globally," people picked it up and passed it around as a nice-sounding, vaguely inspirational slogan—but imagined it was advice they could take or leave alone. Now we are finding out that you really can't get away with *not* thinking globally. The Dubos motto becomes the guideline for our time and—at the very moment of its widest acceptance—also reveals its own obsolescence. The greater truth now emerging in its wake is that the local-global distinction no longer means all that much—not in a world with people on the move, money and goods and information flowing freely, worries about such matters as global climate change forcing their way into our daily thoughts. We simultaneously act locally and globally, think locally and globally, and are hard-put to find an action or an idea that is either purely local or purely global.

It has become fashionable over the past couple of decades to say that we have entered into an "era of limits" and must learn to live within a whole range of new (or newly discovered) constraints such as finite resources and the ability of ecosystems to absorb pollutants. There is some truth in this, but the situation has been enormously misunderstood. Certainly we do not have fewer options or less power, either individually or collectively. On the contrary, we have far more. Each encounter with limits generates more choices, leads to the discovery of new powers and the need for proactive responses. So the world is not so much bumping up against a wall as progressing into a period—to which we can see no end—of continual redefinition of the possible. With technological change and new information, the human species enters new terrain: not once, but over and over.

Proper application of the information to the problems becomes the central challenge of our time. It does not justify a simple faith in an easy "technological fix," nor does it justify the equally simplistic rejection of technology. Instead, it forces people and organizations and governments everywhere to learn as though their lives depended on it—because they do.

And we can see no end point to this learning process. New scientific and technological discoveries seem only to open up more vistas for what is yet to be learned. Nor is there an end point in sight to the learning-about-learning—learning how not to be overwhelmed by information, understanding how to apply it, developing a better sense of the risks and downsides, appreciating the dangers that may result from any use—or non-use—of technology.

Great perils, great possibilities. Much to learn, much to do. A good time to be alive.

Notes

Epigraph

John McHale, *The Future of the Future* (New York: Ballantine, 1969), pp. 107–108.

Chapter One • The Computer Meets the Gene

1. "Kay + Hillis," *Wired*, January 1994, p. 105.
2. Pierre Teilhard de Chardin, *The Phenomenon of Man* (New York: Harper & Row, 1959), p. 219.
3. See Eors Szathmary and John Maynard Smith, "The major evolutionary transitions," *Nature*, March 16, 1995, pp. 227–232.
4. Quoted in Robert Shapiro, *The Human Blueprint: The Race to Unlock the Secrets of Our Genetic Code* (New York: Bantam, 1991), p. 52.
5. Horace Freeland Judson, *The Eighth Day of Creation: The Makers of the Revolution in Biology* (New York: Simon & Schuster, 1979), p. 288.
6. M. Mitchell Waldrop, *Complexity: The Emerging Science at the Edge of the Order and Chaos* (New York: Simon & Schuster, 1992), p. 31.
8. University of Texas System, Center for High Performance Computing, 10100 Burnet Road, Austin, Texas 78758–4497.
9. Stephen Hart, "Test-tube survival of the molecularly fit," *BioScience*, Dec. 1993, p. 741.
10. Gina Kolata, "A.M.A. Is Heralding a Revolution in Genetics," *The New York Times*, Nov. 17, 1993, p. B7.

Chapter Two • Microbionics: A Gathering of Revolutionaries

1. Gordon Rattray Taylor, *The Biological Time Bomb* (New York: New American Library, 1968), p. 13.
2. Elizabeth Antebi and David Fishlock, *Biotechnology: Strategies for Life* (Cambridge: MIT Press, 1986), p. 69.
3. Congress of the United States, Office of Technology Assessment, *Commercial Biotechnology: An International Analysis* (Washington, D.C.: U.S. Government Printing Office, 1984), pp. 143–144.
4. Andrew Pollack, "Genetic Engineers Prepare to Create Brand New Proteins," *The New York Times*, March 15, 1988, p. C1.
5. Peter Simon, "British Drug Giant Glaxo to Buy Affymax," *San Francisco Chronicle*, Jan. 27, 1995, p. D1.

6. Leon Jaroff, "Battler for Gene Therapy," *Time,* Jan. 17, 1994, p. 56.
7. Larry Thompson, *Correcting the Code: Inventing the Genetic Cure for the Human Body* (New York: Simon & Schuster, 1994), p. 32.
8. Gina Kolata, "Novel Bypass Method: A Dose of New Genes," *The New York Times,* December 13, 1994.
9. Tabitha M. Powledge, "The Genetic Fabric of Human Behavior," *BioScience,* June 1993, p. 362.
10. Charles Petit, "Cauliflower Starring in Genetics Lab," *San Francisco Chronicle,* Jan. 27, 1995, p. A5.
11. "Editorial: Toward the Greening of Industry," *EBIS Newsletter* (Brussels: Commission of the European Communities), Vol. 3, no. 2 (May 1993), p. 1.
12. Nathan Rosenberg, "Inventions: Their Unfathomable Future," *The New York Times,* August 7, 1994, p. 9.

Chapter Three • Macrobionics: The Whole Wired World

1. Kevin Kelly, *Out of Control: The Rise of Neo-Biological Civilization* (Reading, Mass.: Addison-Wesley, 1994), p. 440.
2. Lilian Trager (University of Wisconsin-Parkside), panel on "Migration and Population," World Academy of Art and Science conference on "The Governance of Diversity," Minneapolis MN, Sept. 29, 1994.
3. Donald N. Michael, *Cybernation: The Silent Conquest* (Santa Barbara: Center for the Study of Democratic Institutions, 1962).
4. D. James Baker, *Planet Earth: The View from Space* (Cambridge: Harvard University Press, 1990), p. 10.
5. Lewis Thomas, *The Lives of a Cell: Notes of a Biology Watcher* (New York: Bantam, 1975), p. 4.
6. Tom Yulsman, "Virtual Earth," *Earth,* March 1994, p. 24.
7. Trager.
8. Mark Jaffe, *And No Birds Sing: The Story of an Ecological Disaster in a Tropical Paradise* (New York: Simon & Schuster, 1994).
9. William K. Stevens, "American Box Turtles Decline, Perishing Cruelly in Foreign Lands," *The New York Times,* May 10, 1994. p. B5.
10. Alan Burdick, "Attack of the Aliens: Florida Tangles With Invasive Species," *The New York Times,* June 6, 1994, p. B8.
11. Donald D. Plucknett, Nigel J. H. Smith, J. T. Williams, and N. Murthi Anishetty, *Gene Banks and the World's Food* (New Jersey: Princeton University Press, 1987), p. 86.

12. Fred Powledge, "The food supply's safety net," *BioScience,* April 1995, p. 241.
13. *Ibid.*, p. 242.
14. Plucknett *et al.*, pp. 86–87.
15. Quoted in Powledge, p. 238.
16. Peter Raven, "Botanists in a fast-moving world," *New Scientist,* October 13, 1990.
17. Marie-Christine Comte, "(Live)stock Options," *Ceres—The FAO Review,* Nov.–Dec. 1991, p. 16.
18. *Ibid.,* p. 17.
19. Pierre Teilhard de Chardin, *The Phenomenon of Man* (New York: Harper & Row, 1959), p. 182.

Chapter Four • Welcome to the Bio-information Society

1. Merlin Donald, *Origins of the Modern Mind: Three Stages in the Origin of Culture and Cognition* (Cambridge: Harvard University Press, 1991), p. 356.
2. Daniel Bell, "The Social Framework of the Information Society," in Tom Forester (ed.), *The Microelectronic Revolution* (Cambridge: MIT Press, 1981), p. 501.
3. Peter Robinson, *Paul Romer,* Forbes ASAP.
4. The expression "the informatization of society" was used by the French writers Simon Nora and Alan Minc in their book *L'Informatisation de la Societe,* which was unfortunately translated into *The Computerization of Society* when published in the United States. (Cambridge: MIT Press, 1981).
5. Harlan Cleveland, "The Informatization of World Affairs," keynote address, "Bretton Woods Revisited," conference sponsored by The Institute for Agriculture and Trade Policy, Mount Washington Hotel, Bretton Woods, New Hampshire, Oct. 16, 1994.
6. Donald Michael, *On Learning to Plan and Planning to Learn: The Social Psychology of Changing Toward Future-Responsive Social Learning* (Alexandria, Va.: Miles River Press, 1996).
7. Steven A. Rosell *et al., Governing in an Information Society* (Montreal: Institute for Research on Public Policy, 1992), pp. 18–19.
8. Howard Rheingold, "PARC Is Back," *Wired,* Feb. 1994, p. 92.
9. Brenda Fowler, "Recreating Stone Tools to Learn Makers' Ways," *The New York Times,* Dec. 20, 1994, p. C1.

10. Marshall McLuhan, *Understanding Media: The Extensions of Man* (New York: McGraw-Hill, 1965), p. 123.
11. Richard Dawkins, *The Extended Phenotype* (New York: Oxford University Press, 1982).
12. Donald.
13. Henri Bergson, *Creative Evolution,* trans. Arthur Mitchell (London: Methuen, 1954). (First published 1907.)
14. See Thomas Landon Thorson, *Biopolitics* (New York: Holt, Rinehart and Winston, 1970).
15. Donald, p. 279, citing R. Harris, *The Origin of Writing* (London: Duckworth, 1986).
16. Donald, p. 17.
17. Alex Haley, *Roots* (Garden City, New York: Doubleday, 1976), pp. 674–675.
18. For a sampling of the arguments in this debate, see Jack Copeland, *Artificial Intelligence: A Philosophical Introduction* (Oxford: Basil Blackwell, 1983); and the *Daedalus* issue on "Artificial Intelligence," Winter 1988.
19. Donald, p. 382.
20. Bruce Mazlish, *The Fourth Discontinuity: The Co-Evolution of Humans and Machines* (New Haven: Yale University Press, 1993), pp. 6, 233.
21. This is an allusion to Sigmund Freud's famous discussion of the successive blows to the human ego in his *Introductory Lectures to Psychoanalysis.* Mazlish is citing an interpretation of that discussion presented by Jerome Bruner in his "Freud and the Image of Man," *Partisan Review* we, no. 3 (Summer 1956), pp. 340–347.
22. Gregory Stock, *Metaman: The Merging of Humans and Machines into a Global Superorganism* (New York: Simon & Schuster, 1993), p. 20.
23. Richard Rorty, *Contingency, Irony, and Solidarity* (Cambridge University Press, 1989), p. 73.
24. Peter F. Drucker, *Managing for the Future: The 1990s and Beyond* (New York: Dutton, 1992), p. 329.
25. Harlan Cleveland, *The Knowledge Executive: Leadership in an Information Society* (New York: E. P. Dutton, 1985), pp. 22–23.

Chapter Five • Augmentations Old and New

1. Quoted in Janice M. Cauwels, *The Body Shop: Bionic Revolutions in Medicine* (St. Louis: C. V. Mosby Company, 1986), pp. 285–286.
2. Allan Chase, *Magic Shots* (New York: William Morrow, 1982), p. 42.
3. William H. McNeill, *Plagues and Peoples* (New York: Doubleday, 1977), p. 235.
4. Chase, p. 46.
5. Florence Nightingale, *Notes on Nursing* (1859), (New York: Dover, 1969), pp. 32–33n.
6. Herbert Spencer, *The Study of Sociology* (1874), quoted in Chase, p. 68
7. Lawrence K. Altman, "Frozen in Labs, Smallpot Virus Escapes a Scalding End, for Now," *The New York Times,* Dec. 25, 1993, p. 1.
8. Lawrence K. Altman, "After Long Debate, Vaccine for Chicken Pox Is Approved," *The New York Times,* March 18, 1995, p. 1.
9. Tore Godal, quoted in John Maurice, "Malaria Vaccine Raises a Dilemma," *Science,* Jan. 20, 1995, p. 323.
10. Chiron 1994 Annual Report, *Building Value Through Cooperation,* p. 13. Chiron Corporation, 4560 Horton Street, Emeryville CA 94608-2916.
11. Cauwels, p. 17.
12. Malcolm W. Browne, "Researchers Develop a Bone Healer, but a Patent Issue Arises," *The New York Times,* March 25, 1995, p. 7.
13. Test-Tube Cartilage Is Found to Repair Knees," *The New York Times,* Sept. 6, 1994, p. A9.
14. Jonathan D. Beard, "Fresh Approaches to a Familiar Problem: Skin Replacement," *The New York Times,* Jan. 30, 1994, p. 9.
15. Stephen J. Benkovic (Department of Chemistry, Pennsylvania State University), in "Through the Glass Lightly," *Science,* March 17, 1995, p. 1618.
16. Mark Dowie, *"We Have a Donor": The Bold New World of Organ Transplanting* (New York: St. Martin's Press, 1988), p. 74.
17. Malcolm W. Browne, "How Brain Waves Can Fly a Plane," *The New York Times,* March 7, 1995, p. B5.
18. Barbara B. Brown, *Stress and the Art of Biofeedback* (New York: Bantam, 1978), p. 3.

19. Mark S. Schwartz and associates, *Biofeedback: A Practitioner's Guide* (New York: The Guilford Press, 1987), p. 506.
20. Peter F. Drucker, *The Age of Discontinuity: Guidelines to Our Changing Society* (New York: Harper & Row, 1968).
21. Harlan Cleveland, *The Knowledge Executive* (New York: E. P. Dutton, 1985).
22. Joe Flower, "The Other Revolution in Health Care," *Wired,* Jan. 1994, p. 110.

Chapter Six • The Human–Animal Connection

1. Leah Garchik, "Homage to Senator Helms," *San Francisco Chronicle,* March 9, 1995, p. D12. (Sen. Helms interviewed on TV by John McLaughlin.)
2. William H. Allen, "Farming for Spare Body Parts," *BioScience,* February 1995, p. 73.
3. *Ibid.,* p. 74.
4. U.S. Congress, Office of Technology Assessment, *Alternatives to Animal Use in Research, Testing and Education* (Washington, D.C.: U.S. Government Printing Office, Feb. 1986).
5. Bernard E. Rollin, *Animal Rights and Human Morality* (Buffalo, N.Y.: Prometheus, 1981), p. 91.
6. M. Windeatt and W. May, "Charles River: The General Motors of Animal Breeding," *The Beast,* Summer 1980, p. 27.
7. W. M. S. Russell and R. L. Burch, *The Principles of Humane Experimental Technique* (London: Methuen, 1959).
8. Institute for Laboratory Animal Resources, *National Survey of Laboratory Animal Facilities and Resources.* (Washington, D.C.: U.S. Department of Health and Human Services, NIH Publication No. 80-2091, 1980).
9. Peter Singer, *Animal Liberation: A New Ethics for Our Treatment of Animals* (New York: Avon, 1975), p. 48. Also see John Parascandola, "The Development of the Draize Test for Eye Toxicity," *Pharmacy in History,* Vol. 33 (1991) No. 3.
10. Charlene Crabb, "Tinkering with Rodents to Probe Heredity's Mysteries," *U.S. News & World Report,* November 4, 1991, p. 70.
11. Lawrence M. Fisher, "Athena Neurosciences Makes Itself Heard in Battle Against Alzheimer's," *The New York Times,* Feb. 15, 1995, p. C4.

12. Richard D. French, *Antivivisectionism and Medical Science in Victorian Society* (Princeton, N.J.: Princeton University Press, 1975); James Turner, *Reckoning with the Beast: Animals, Pain and Human ity in the Victorian Mind* (Baltimore, Md.: Johns Hopkins University Press, 1980).
13. Michael Allen Fox, *The Case for Animal Experimentation: An Evolutionary and Ethical Perspective* (Berkeley: University of California Press, 1986), p. 184.
14. Michael W. Fox, *Inhumane Society: The American Way of Exploiting Animals* (New York: St. Martin's Press, 1990), pp. 72–73.
15. William Paton, *Man and Mouse: Animals in Medical Research,* (New York: Oxford University Press, 1984).
16. California Biomedical Research Association, 1008 Tenth St., Suite 328, Sacramento, CA 95814.
17. Peter Singer, "Ten Years of Animal Liberation," *The New York Review of Books,* Jan. 17, 1985, p. 8.
18. Andrew Rowan, *Of Mice, Models and Men: A Critical Evaluation of Animal Research* (Albany, N.Y.: SUNY Press, 1984), p. 181.
19. Hans Ruesch, *Slaughter of the Innocent* (New York: Bantam, 1987), p. 225.
20. Steven Levy, *Artificial Life: A Report from the Frontier Where Computers Meet Biology* (New York: Vintage, 1992).
21. Percy B. Medawar, *The Hope of Progress* (London: Methuen, 1972), p. 86.

Chapter Seven • The Body Politic: Private Lives and Public Issues

1. Jeffrey A. Fisher, *Our Medical Future: Breakthroughs in Health and Longevity by the Year 2000 and Beyond* (New York: Pocket Books, 1992), p. 24.
2. Mark Dowie, *"We Have a Donor," The Bold New World of Organ Transplanting* (New York: St. Martin's Press, 1988).
3. John Henkel, "Safeguarding Human Tissue Transplants," *FDA Consumer,* September 1994, p. 10.
4. Dr. Donald Laub, quoted in Amy Bloom, "A Reporter at Large: The Body Lies," *The New Yorker,* July 18, 1994, p. 47.
5. Gina Kolata, "Doomed Babies Are Seen As the Donors of Organs," *The New York Times,* May 24, 1995, p. B6.

6. D. O. Cauldwell, "Psychopathia transsexualis," *Sexology,* no. 16, 1949, pp. 274–280.

7. Jon K. Meyer and John E. Hoopes, "The Gender Dysphoria Syndrome," *Newsweek,* Nov. 22, 1976, cited in Janice G. Raymond, *The Transsexual Empire* (Boston: Beacon Press, 1979), p. 23.

9. Bloom, p. 44.

10. Anne Fausto-Sterling, *Myths of Gender: Biological Theories About Women and Men* (New York: Basic Books, 1992).

11. "China's coming 'Marriage Gap': A Million Brideless Men a Year by 2020," *Stanford News,* Feb. 8, 1995. (Stanford University News Service.)

12. Fisher.

13. Laura Bird, "Birth-Control Market Sees Barrage of Ads," *The Wall Street Journal,* Nov. 11, 1993, p. B10.

14. William D. Montalbano, "Italy's Doctors Ban 'Designer Babies,'" *San Francisco Examiner,* April 16, 1995, p. A-5.

15. Montalbano, *Ibid.*

16. Fisher.

17. Untitled information brochure, Dor Yeshorim, Committee for Prevention of Jewish Genetic Diseases, Brooklyn, New York.

18. Gina Kolata, "Nightmare of the Dream of a New Era in Genetics?" *The New York Times,* Dec. 7, 1993, pp. A1–B8.

19. Diane B. Paul and Hamish G. Spencer, "The hidden science of eugenics," *Nature,* March 23, 1955, p. 302.

20. Steve Jones, "Our Genetic Future: The Evolution of Utopia," *The (London) Independent,* Dec. 19, 1991, p. 12.

21. Arthur L. Caplan, in D. Bartels (ed.), *Prescribing Our Future: Ethical Challenges in Genetic Counselling* (Hawthorne, N.Y.: Aldine de Gruyter, 1993), pp. 149–165.

22. Ethical Issues Policy Statement on Huntington's Disease Molecular Genetics Productive Test, *Journal of Molecular Genetics,* 27, 1990, pp. 34–38.

23. Jones, *Ibid.*

24. Paul H. Silverman, "Human Germ-line Gene Alteration: An Approaching Reality," unpublished draft manuscript.

25. Jones, *Ibid.*

26. "Points to consider in the design and submission of protocols

for the transfer of recombinant DNA into the genome of human subjects" as approved by Human Gene Therapy subcommittee and the NIH Recombinant DNA Advisory Committee (RAC). *Human Gene Therapy,* no. 1, 1990, pp. 93–103. Also see Rebecca Kolberg, "RAC tiptoes into new territory: In utero gene therapy," *The Journal of NIH Research,* no. 7, Jan. 1995. pp. 37–39.

27. Silverman.
28. Lawrence M. Fisher, "The Stuff of Dreams Nears Reality," *The New York Times,* June 1, 1995, p. C1.

Chapter Eight • Reinventing Agriculture

1. Lawrence Busch, William B. Lacy, Jeffrey Burkhardt, and Laura R. Lacy, *Plants, Power, and Profit: Social, Economic, and Ethical Consequences of the New Biotechnologies* (Cambridge, Mass.: Blackwell, 1991), p. 33.
2. William K. Stevens, "Dry Climate May Have Forced Invention of Agriculture," *The New York Times,* April 2, 1991, p. B5.
3. Pat Roy Mooney, *Seeds of the Earth: A Public or Private Resource?* Ottawa: Canadian Council for International Cooperation, 1979), p. 11.
4. Edward C. Wolf, *Beyond the Green Revolution: New Approaches for Third World Agriculture* (Washington, D.C.: Worldwatch Institute, 1986).
5. Robert Walgate, *Miracle or Menace? Biotechnology and the Third World.* (London: The Panos Institute, 1990), p. 46.
6. Keith Griffin, *The Political Economy of Agrarian Change: An Essay on the Green Revolution* (Cambridge, Mass: Harvard University Press, 1974), p. xiii.
7. See, for example, Busch *et al.,* p. 51.
8. Gar Smith, "Man-made Bacteria Halted in the Fields: Secretly Released in Oakland," *Earth Island Journal,* March 1986, p. 4.
9. Gregg Easterbrook, *A Moment on the Earth* (New York: Viking, 1995), p. 119.
10. Lawrence M. Fisher, "Monsanto to Acquire 49.9% of Biotechnology Company," *The New York Times,* June 29, 1995, p. C3.
11. Sibella Kraus, "Brave New Vegetables," *San Francisco Examiner,* July 8, 1992, food section, p. 1.
12. Steve Rhodes, "Friendly Farming: How Green Is My Acre," *Newsweek,* Dec. 26, 1994, p. 113.

13. Fourth Annual Report (1993–1994), M. S. Swaminathan Research Center (Madras). See also M. S. Swaminathan (ed.), *Ecotechnology and Rural Development: A Dialogue* (Madras: Macmillan India Ltd., 1994).
14. A. S. Bhalla and Dilmus James (eds.), *New Technologies and Development: Experiences in "Technology Blending"* (Boulder, Co.: Lynne Rienner, 1988).
15. Steve Tally, "2001: A Farm Odyssey," undated publication, Department of Agricultural Communications, Purdue University.
16. Walter Truett Anderson, "Food without Farms: The Biotech Revolution in Agriculture," *The Futurist*, January–February 1990, p. 21.
17. Benjamin Pogrund, "Wresting food from concrete jungles," *The World Paper*, June 1995, p. 10.
18. Nancy Jack Todd and John Todd, *Bioshelters, Ocean Arks, City Farming: Ecology as the Basis of Design* (San Francisco: Sierra Club Books, 1984).
19. Alex Barnum, "Making Medicines in Biotech Barns," *San Francisco Chronicle*, July 6, 1993, p. E1.
20. Busch *et al.*, pp. 27–30.

Chapter Nine • Bio-info Industries

1. George Steiner, "Life-Lines," *The New Yorker*, March 6, 1971, p. 101.
2. Ronald M. Atlas and Carl E. Cerniglia, "Bioremediation of Petroleum Pollutants," *BioScience*, May 1995, p. 332.
3. Karen Bernstein, "Do Microbes Hold the Key to Toxic Waste Clean-up? The EPA Thinks so," *BioWorld*, Nov./Dec. 1990, p. 46.
4. Glenn Zorpette, "Food Indigo," *Scientific American*, July 1995, p. 29.
5. "Uranium-hungry microbes filter toxic wastes," *Genetic Engineering and Biotechnology Monitor*, November 1991, p. 39.
6. *Hazardous Materials: Microbiological Decomposition* (Springfield, Va.: National Technical Information Service, 1992).
7. Richard C. Cassin of Bioremediation, Inc. (San Diego, Calif.), quoted in Robert D. Hof, "The Tiniest Toxic Avengers," *Business Week*, June 4, 1990, p. 96.
8. Elizabeth Antebi and David Fishlock, *Biotechnology: Strategies for Life* (MIT Press, 1986), pp. 153–154.

9. Carl-Goran Heden, "Potential of Biotechnology," address to commemorate the establishment of the United Nations University Institute on New Technologies (INTEC), Maastricht, The Netherlands, June 28, 1990.
10. Bonnie Wyper, "Nature: A Material World," *What Next?*, April 1995, p. 12.
11. Quoted in Barnaby J. Feder, "Farmers' Crops Are Holding Promise Beyond Food," *The New York Times*, Jan. 25, 1995, p. C6.
12. Amy O'Marro, "The Emergence of EcoPLA," *Cargill News*, Minneapolis, Minn., Sept. 1993.
13. Charles Petit, "Gene-Altered Plants Produce Plastic," *San Francisco Chronicle*, Feb. 27, 1995, p. 1.
14. Christopher Flavin and Nicholas Lenssen, "The Unexpected Rise of Natural Gas," *The Futurist*, May–June 1995, p. 34.
15. Nigel Calder, *The Green Machines* (New York: Putnam's, 1986), p. 45.
16. Johanna Dobereiner, "The Brazilian Biofuel Programme: Solution to Social and Environmental Problems," in M. S. Swaminathan (ed.), *Ecotechnology and rural Employment* (Madras: Macmillan India, 1994), p. 121.
17. Neil Gross, "The Green Giant? It May Be Japan," *Business Week*, Feb. 24, 1992, p. 74.
18. Robert R. Birge, "Protein-Based Computers," *Scientific American*, March 1995, p. 90.
19. Leonard M. Adleman, "Molecular Computation of Solutions to Combinatorial Problems," *Science*, Nov. 11, 1994, pp. 1021–1024.
20. Gina Kolata, "A Vat of DNA May Become Fast Computer of the Future," *The New York Times*, April 11, 1995, p. B8.
21. Robert Socolow, Princeton University Center for Energy and Environmental Studies, as quoted in Gregg Easterbrook, *A Moment on the Earth* (New York: Viking, 1995), pp. 360–361.

Chapter Ten • Human Governance of Natural Biosystems

1. James E. Lovelock, *Gaia: A New Look at Life on Earth* (Oxford University Press, 1979), pp. 61–62.
2. Peter Raven, "Botanists in a fast-moving world," *New Scientist*, Oct. 13, 1990, p. 45.

3. Norbert Wiener, *Cybernetics: Or Control and Communication in the Animal and the Machine* (New York: Wiley, 1948).
4. David Binder, "For Moose No. 9, a 600-Mile Checkup," *The New York Times,* March 22, 1994, p. 35.
5. David Foster (Associated Press), "Biologists Returning wolves to Rockies," *San Francisco Examiner,* Jan. 8, 1995, p. B-8.
6. Newsletter, Center for Biotechnology Policy and Ethics, Texas A&M University, Jan. 4, 1992, p. 4.
7. Catharine Skipp, "Cougars Enlisted in Effort to Save the Florida Panther," *The New York Times,* May 11, 1993, p. B6.
8. Catherine Dold, "Florida Panthers Get Some Outside Genes," *The New York Times,* June 20, 1995, p. B7.
9. Peter Passell, "One Answer to Overfishing: Privatize the Fisheries," *The New York Times,* May 11, 1995.
10. Malcolm W. Browne, "Rabies, Rampant in U.S., Yields to Vaccine in Europe," *The New York Times,* July 5, 1994, p. B5.
11. Michael Hill (*Baltimore Sun*), "Neighbors Covet South African Game Park," *San Francisco Chronicle,* May 30, 1995, p. A7.
12. See Micah Morrison, *Fire in Paradise: The Yellowstone Fires and the Politics of Environmentalism* (New York: HarperCollins, 1993).
13. George J. Mitchell, *World on Fire: Saving an Endangered Earth.* (New York: Scribner's, 1991), pp. 70–71.
14. Thomas Gale Moore, *Global Warming: A Boon to Humans and Other Animals* (Stanford University: Hoover Institution, 1995).
15. Sonja A. Boehmer-Christiansen, "A Scientific Agenda for Climate Policy?" *Nature,* Dec. 1994, pp. 400–402.
16. Paul Wapner, "On the Global Dimension of Environmental Challenges," *Politics and the Life Sciences,* August 1994, p. 180.
17. *Ibid.,* p. 173.
18. Mark Cherrington, "Weather or Not," *Earthwatch,* May/June 1995, p. 5.
19. David Perlman, "Scientists Warn of Asteroids," *San Francisco Chronicle,* May 23, 1995, p. 1.
20. Murray Gell-Mann, *The Quark and the Jaguar* (New York: W. H. Freeman, 1994), p. 17.
21. *Ibid.,* p. 73.

22. My summary is based on M. Mitchell Waldrop, *Complexity: The Emerging Science at the Edge of Order and Chaos* (New York: Simon & Schuster, 1992), pp. 145–147. See also Philip W. Anderson, Kenneth J. Arrow, and Daniel Pines (eds.), *The Economy as an Evolving Complex System,* Santa Fe Institute Studies in the Sciences of Complexity, vol. 5. (Redwood City, Calif.: Addison-Wesley, 1988).
23. Ilya Prigogine and Isabelle Stengers, *Order out of Chaos: Man's New Dialogue with Nature* (New York: Bantam Books, 1984).

Chapter Eleven • Changing Shades of Green

1. Gregg Easterbrook, *A Moment on the Earth: The Coming Age of Environmental Optimism* (New York: Viking, 1995), p. 369.
2. Daniel B. Botkin, *Discordant Harmonies: A New Ecology for the Twenty-First Century* (New York: Oxford University Press, 1990), p. 193.
3. George Perkins Marsh, *Man and Nature, or, Physical Geography as Modified by Human Action* (Cambridge: Harvard University Press, 1965). (First published 1864.)
4. Walt Anderson, *Politics and Environment: A Reader in Ecological Crisis* (Pacific Palisades, Calif.: Goodyear/Prentice Hall, 1971). Second edition (rev.) 1976.
5. Thomas Berry, *The Dream of the Earth* (San Francisco: Sierra Club Books, 1988), p. 82.
6. Easterbrook, p. xvi.
7. *Ibid.,* p. 648.
8. Don Ogden, "The Longest War: Rooted in 'Civilization'?" *Peacework* (American Friends Service Committee), April 1995, p. 24.
9. Kirkpatrick Sale, *Dwellers in the Land: The Bioregional Vision* (San Francisco: Sierra Club books, 1985), p. 54
10. *Ibid., p. 116.*
11. Bill Devall and George Sessions, *Deep Ecology: Living As If Nature Mattered* (Salt Lake City: Peregrine Smith Books, 1985), p. 148.
12. Daniel Conner, "Is AIDS the Answer to an Environmentalist's Prayer?" *Earth First!,* Dec. 23, 1987, pp. 14–16.
13. Martin W. Lewis, *Green Delusions: An Environmentalist Critique of Environmental Radicalism* (Durham, NC: Duke University Press, 1992), p. 11.
14. Marsh, pp. 29–30.

15. Herman Daly, *Steady-State Economics: The Economics of Biophysical Equilibrium* (San Francisco: W. H. Freeman, 1977).
16. Botkin, p. 9.
17. Botkin, pp. 58–59.
18. Elizabeth Culotta, "Bringing Back the Everglades," *Science*, June 23, 1995, p. 1689.
19. Robert K. Colwell, Lawrence Barnthouse, Andrew Dobson, Frieda Taub, and Richard Wetzler, letter to U.S. Office of Science and Technology Policy, Sept. 20, 1986.

Chapter Twelve • The Infinite Schoolhouse

1. Quoted in John Horgan, "Perpendicular to the Mainstream," *Scientific American*, Aug. 1993, p. 27.
2. See, for example, Ewert H. Cousins (ed.), *Process Theology: Basic Writings by the Key Thinkers of a Major Modern Movement* (New York: Newman Press, 1971).
3. Richard Dawkins, *The Blind Watchmaker: Why the Evidence of Evolution Reveals a Universe Without Design* (New York: Norton, 1987).
4. Susantha Goonatilake, *The Evolution of Information: Lineages in Gene, Culture and Artefact* (London: Pinter, 1991).
5. See Eric Jantsch and Conrad H. Waddington (eds.), *Evolution and Consciousness* (Reading, Mass.: Addison-Wesley, 1976). Also George Gaylord Simpson, *The Meaning of Evolution* (rev. ed) (New Haven, Conn.: Yale University Press, 1967), p. 250.
6. See Daniel C. Dennett, *Darwin's Dangerous Idea* (New York: Simon & Schuster, 1995).
7. Susantha Goonatilake, "The New Technologies and the 'End of History,'" *Futures Research Quarterly*, Summer 1993, p. 85.
8. Walter Truett Anderson, *Reality Isn't What It Used to Be* (San Francisco: HarperCollins, 1990). Also see *The Truth about the Truth: De-confusing and Re-constructing the Postmodern World* (New York: Tarcher/Putnam, 1995).
9. The significance of headhunting is discussed in Michelle Z. Rosaldo, *Knowledge and Passion* (Cambridge: Cambridge University Press, 1980), and cited as an example of a cultural construction in Kenneth J. Gergen, *The Saturated Self: Dilemmas of Identity in Contemporary Life* (New York: Basic Books, 1991). Ethnographic reports from New Guinea are cited in Richard Shweder, "Why Do Men

Barbecue? and Other Postmodern Ironies of Growing Up in the Decade of Ethnicity," *Daedalus,* Winter 1993, Vol. 122, Number 1.

10. Bill McKibben, *The End of Nature* (New York: Random House, 1989).
11. Harold Lasswell, *Politics: Who Gets What, When, How* (New York: Peter Smith, 1950).
12. Quoted in "Technology Transfer," European Biotechnology Information Service Newsletter, Dec. 1993, p. 5.
13. Stephen S. Hall, "IL-12 at the Crossroads," *Science,* June 9, 1995, p. 1434.
14. Wilfred Beckerman, *Small Is Stupid: Blowing the Whistle on the Greens* (London: Duckworth, 1995).
15. Kevin Kelly, *Out of Control: The Rise of Neo-Biological Civilization* (Menlo Park, Calif.: Addison-Wesley, 1994), p. 469.
16. Walter Wriston, *The Twilight of Sovereignty: How the Information Revolution Is Transforming Our World* (New York: Scribner's, 1992), p. 45.
17. Daniel Dayan and Elihu Katz, *Media Events: The Live Broadcasting of History* (Cambridge: Harvard University Press, 1992), p. 26.
18. Henry Kissinger, *Diplomacy* (New York: Simon & Schuster, 1994).
19. See James N. Rosenau, *Turbulence in World Politics: A Theory of Change and Continuity* (N.J.: Princeton University Press, 1990).
20. See Harlan Cleveland, *Birth of a New World: An Open Moment for International Leadership* (San Francisco: Jossey-Bass, 1993).
21. See Mary Catherine Bateson, "Toward an Ambiguous World Order," in Richard A. Falk, Robert C. Johansen, and Samuel S. Kim (eds.), *The Constitutional Foundations of World Peace* (Albany, N.Y.: SUNY Press, 1993).

Index